H. Lippert/R. Pabst · Arterial Variations in Man

Herbert Lippert and Reinhard Pabst

Arterial Variations in Man

Classification and Frequency

With 540 Drawings in 90 Two-color-plates

J. F. Bergmann Verlag München

Professor Dr. med. Dr. phil. H. Lippert
Zentrum Anatomie
Medizinische Hochschule Hannover
Abteilung für Funktionelle und Angewandte Anatomie
Postfach 61 01 80
3000 Hannover 61

Professor Dr. med. R. Pabst
Zentrum Anatomie
Medizinische Hochschule Hannover
Abteilung für Topographische Anatomie und Biomechanik
Postfach 61 01 80
3000 Hannover 61

ISBN-13:978-3-642-80510-3 e-ISBN-13:978-3-642-80508-0
DOI: 10.1007/978-3-642-80508-0

Softcover reprint of the hardcover 1st edition 1985

Contents

1 Introduction

In textbooks on anatomy, radiology and surgery only the "normal" arterial blood supply is usually described. This "normality", however, is sometimes found in less than 30% of all cases for some arteries, but in over 95% for others. Rarely mentioned are deviations in the individual artery's origin, topographical localization and the area it supplies. They can be classified into two groups: malformations and variations. Malformations often have a negative influence on the function of the organ under normal circumstances, e.g. if both coronary arteries originate from the pulmonary artery. This group will be dealt with in only a few cases. Variations, in contrast, generally have no effect on the function of the organ under normal circumstances, e.g. a superficial brachial artery does not have a negative effect on the function of the forearm and hand. However, if the superficial artery were mistaken for a vein and a thiobarbiturate were injected, severe necrosis of the hand would result. Thus, even a harmless variation can have negative effects under certain circumstances.

There has recently been a surge of clinical interest in the topographical anatomy of the arterial blood supply, the origin of arteries and the area supplied by individual vessels. Superselective angiography can now be performed on many organs, and smaller arteries and branches, which were formerly identified only in carefully dissected anatomical preparations, can now be visualized in vivo. Modern surgical techniques depend in part on knowledge of both the "normal" and the anomalous arterial blood supply. For instance, when selective arterial chemotherapy is used for an hepatic cancer, aberrant hepatic arteries can pose a significant problem. Modern microsurgical techniques employed in vascularized transplants and repairs after trauma are based on a sound knowledge of the variations of arteries.

Many descriptive terms are found in the literature such as aberrant, replaced, supplementary or accessory arteries. We have used the term "replaced artery" to refer to a single artery which supplies an organ instead of the normal one. An "accessory artery" is a second artery in addition to the one normally present, without any specification of size being made. However, there is no general agreement on whether minute vessels with very small diameters and hardly any significant blood flow should also be considered.

The aims of this book were twofold: first, to extract the frequency of arterial anomalies from the literature (often published in inaccessible journals) and second, to classify these arteries by schematic outlines of the basic types.

The determination of their frequency poses some obvious problems. The numerous descriptions of single cases are of little use for estimating frequency, but are very important for demonstrating the existing types of anomalies. In general, there have been two kinds of studies:

A. **Angiologic data** are never based on a representative group of patients, since obviously there was an indication for the angiograph. Smaller variant arteries might be missed in selective arteriography or simply overlooked because they were not expected. In certain diseases small branches increase in size, thus making it difficult to define whether they are variants of the normal blood supply or represent a pathological condition.

B. In **anatomical studies** the frequencies of variant arteries are sometimes lower than in reality, because small accessory vessels can be missed or cut, and not all arteries are filled in corrosion cast preparations. Variant arteries that produce clinically significant effects have been studied much more frequently by angiography or dissected in large numbers of corpses. New surgical techniques are now directing attention to other arteries. Due to the increasing numbers of coronary angiographies and aortocoronary anastomoses performed many papers have been published on the varieties of coronary arteries. This could foster the false impression that most had not been described previously. Finally, although racial differences may play a role, the paucity of data on this aspect does not permit it to be covered here.

The classification of arterial variations is usually based on the normal embryological development. Rapid growth occurs during ontogeny, with anastomoses between arteries disappearing. However, some of these arteries, usually only present for a short period, remain throughout life. Furthermore, as many organs, such as the testes and the heart, wander during their development, a knowledge of their original localization may explain certain abnormalities. No detailed descriptions of the different types are given, only brief explanations. More important are the schematic drawings, which show the variations of the origin, the course of the artery and sometimes the area or portion of the organ supplied. The figures are arranged mainly according to individual arteries or the blood supply of different organs. There will of necessity be some overlap, especially in areas like the celiac trunk. The reader looking for one artery in the index will find all variants of that vessel listed together on a few pages with the corresponding drawings.

The numerous descriptions of single cases of an abnormal artery could not all be cited in the references. Such papers were selected if a rare individual type was of special interest for any clinical reason or to explain the development of variations of that artery in general. Preference was given to articles reporting large numbers of patients studied by angiography or a greater series of autopsies. Some earlier papers were excluded if published in journals not readily accessible or if quoted in more recent papers of a review type. All references were not given for all subtypes, since this would have resulted in much repetition. The references for each chapter are divided into reviews (marked by an asterisk [*]), books, and case reports of a few patients.

We would like to ask readers to inform us of other anomalies known to them but not included in this book and to draw our attention to recent or upcoming papers on anomalies of arteries in man.

This book arose out of a series of schematic drawings of arterial anomalies, which appeared as separate plates in a German medical weekly about 15 years ago. We are grateful to Prof. Dr. H. J. Clemens of Bergmann Verlag, Munich, for encouraging us to publish these figures, augmented by additional ones and including references. Most of the figures were drawn by Mrs. A. Luhmann with great patience and expertise. Figures pp. 3–15, 18–19, 23–24 were drawn by Mrs. N. von Redwitz and Mr. S. Nüssel. We would also like to thank Ms S. Fryk for excellent secretarial assistance.

2 Aortic arch

Development of the aortic arch

During the early stages of embryonic development two pairs of aortas are present. The ventral aortas ascend from the heart, turn dorsally within the first branchial arch and descend as the dorsal aortas. Already in embryos with 3 mm crown-heal length the beginnings and ends of the paired aortas merge, remaining separate only in the area of the foregut. In each of the six branchial arches connections develop between the ventral and dorsal aortas, the branchial arteries. These arteries do not coexist, the first branchial arteries having already disappeared before the fifth and sixth develop. The fifth branchial artery seems to be present for only a few hours, although a few instances of its persisting have been reported [275, 340, 497]. The carotid arteries develop from the cranial part of the ventral and dorsal aortas. The dorsal aortas give off segmental branches along their segmented body wall: 3 occipital, 7 cervical and 12 thoracic, etc. All cervical arteries disappear, except for the sixth which forms the subclavian artery. A longitudinal anastomosis remains on both sides to form the vertebral artery.

Thus, as a rule, the human aortic arch develops in the following way: 1. The left side of the 4th branchial artery forms part of the aortic arch, and the right side forms the beginning of the subclavian artery. 2. Parts of the dorsal aortas on both sides atrophy, i.e., the area between the 3rd and 4th branchial artery (left) and the section between the 6th segmental artery and the merged descending aorta (right). 3. The 6th branchial arteries form the beginning of the pulmonary arteries and the ductus arteriosus

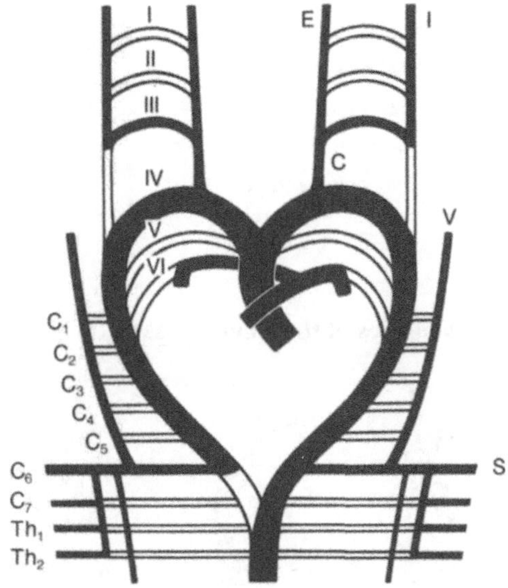

C = common carotid artery
E = external carotid artery
I = internal carotid artery
S = subclavian artery
V = vertebral artery

(Botallo's duct). The final topographical position of the aortic arch and its branches is a product of differential growth rates of various parts of the arteries, which results in a "migration" and "merging" of branches. The main force behind these changes seems to be the optimization of hemodynamic paths combined with the descending heart. [References for developmental and general aspects of the aortic arch: 2, 15, 38*, 61, 126, 145, 148, 150*, 151*, 183, 202, 205*, 233, 234, 372, 561, 589, 567*.]

1. "Normal" situation as given in textbooks ~70%

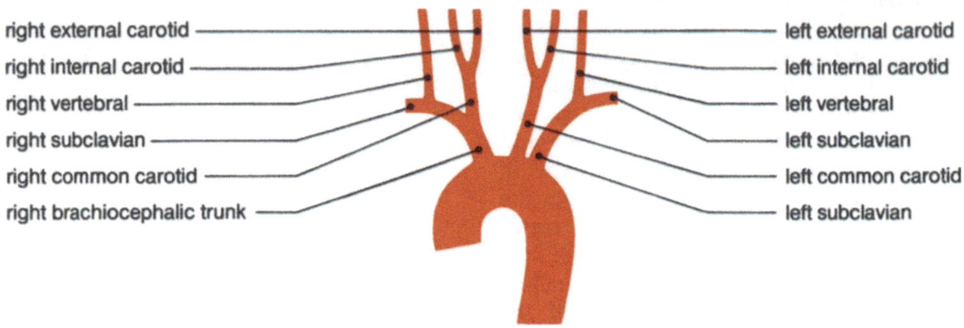

2. Anomalies of the trunk ~23%

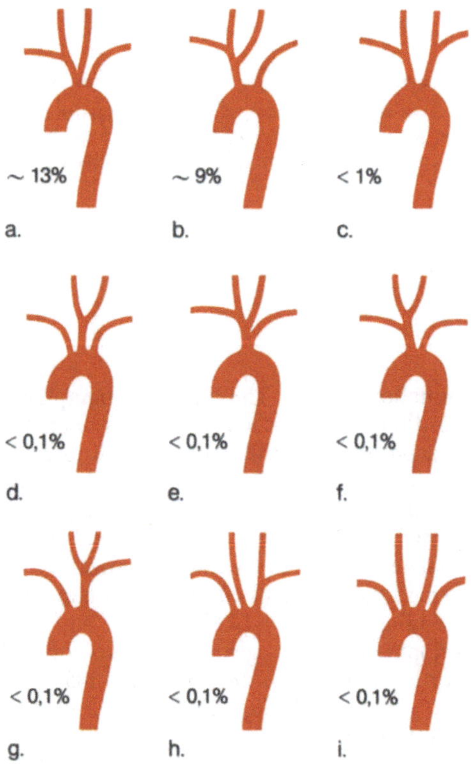

a. common origin of the right brachiocephalic trunk and left common carotid artery
b. left common carotid artery originates from the right brachiocephalic trunk
c. right and left brachiocephalic trunk

d. trunk formation of both carotid arteries (bicarotid)
e. common brachiocephalic trunk
f. right subclavian artery originates from a bicarotid trunk

g. left subclavian artery originates from a bicarotid trunk
h. only a left brachiocephalic trunk
i. no brachiocephalic trunk

The frequencies of the different types depend largely on the method of examination and racial factors (Type 2a and 2b seem to be present more often in Blacks than in Caucasians). Some descriptions are difficult to classify and lie between types 1, 2a and 2b. According to the literature, type 1 + type 2 cover about 93% of all humans [2, 15*, 19, 31, 33, 35, 48, 61, 90, 127, 146, 150*, 179, 257, 306, 320, 355, 513, 527, 567*, 612, 654, 661]. Some types that are considered anomalies in humans are the rule in other mammals, e.g. type 2b occurs in rodents and carnivores, 2c in insectivores, 2d in elephants and 2e in paired and unpaired ungulates [234].

3. Vertebral artery as a direct branch of the aortic arch [4%]

a. left vertebral artery as the penultimate branch of the aortic arch
b. left vertebral artery as the penultimate branch of the aortic arch; left common carotid artery originating from the brachiocephalic trunk
c. left vertebral artery as the last branch of the aortic arch

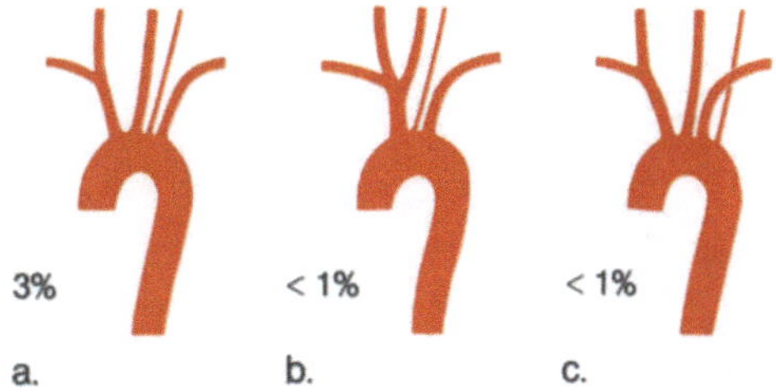

d. left vertebral artery as the last branch of the aortic arch; left common carotid artery originating from the brachiocephalic trunk
e. left vertebral artery as the last branch of the aortic arch; a common brachiocephalic trunk
f. vertebral artery branches before the left subclavian artery, but the last branch of the aortic arch is the right subclavian artery

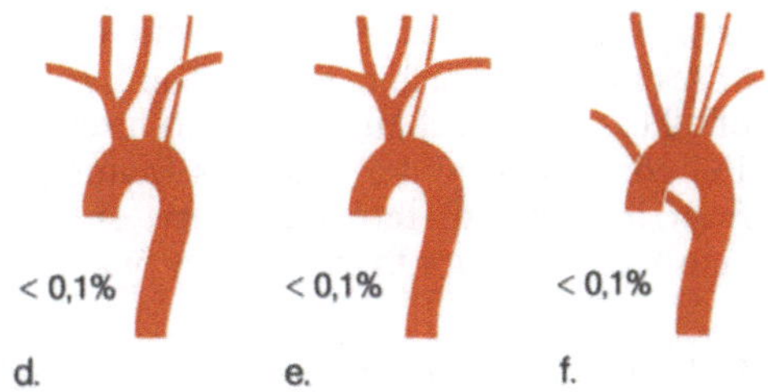

g. two roots of the vertebral artery, one the penultimate branch of the aortic arch
h. both vertebral arteries directly branch from the aortic arch

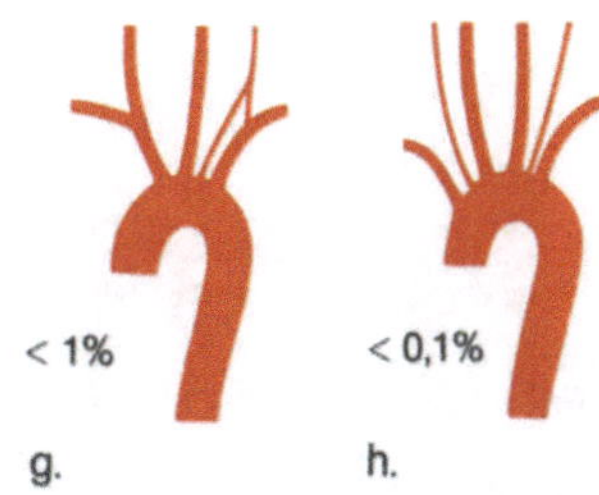

I. development of type 3 a

II. development of type 3 g

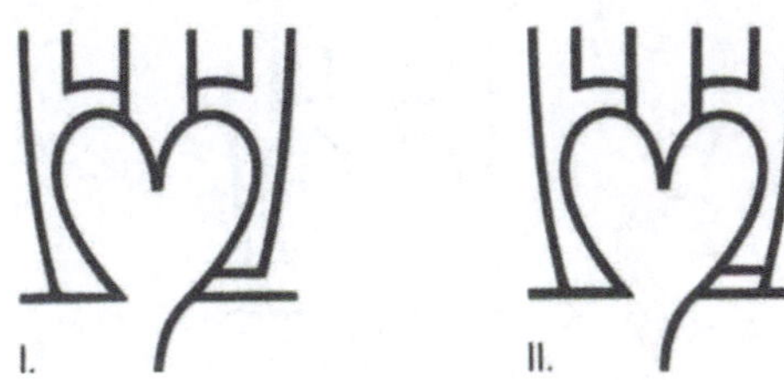

When a segmental artery persists more cranial than the 6th cervical artery, the left vertebral artery will branch from the aortic arch. In such cases the vertebral artery enters the vertebral column through a more cranial transverse foramen. The vertebral artery can have two origins when the longitudinal anastomosis to the 6th segmental artery remains open. In extremely rare instances the right vertebral artery originates from the aortic arch. In such cases either all the beginning part of the right 4th branchial artery forms the aortic arch or there is a variety of type 5 with a subsequent "migration" of the origin of the artery.

4. Arteria thyroidea ima as a direct branch of the aortic arch [1%]

The overall frequency of the lowest thyroid artery (a. thyroidea ima) is about 6%: it arises from the brachiocephalic trunk in 3% of all cases, right common carotid artery in 1%, aortic arch in 1%, and less frequently from the internal thoracic artery, subclavian artery, inferior thyroid artery or suprascapular artery. Two arteria thyroidea ima have also been described. This artery is of importance in operations performed caudal to the isthmus of the thyroid gland. [See references for general features; 7, 15*, 113, 150*, 151*, 247, 253, 389, 412, 432, 506, 516.]

5. Right subclavian artery as the last branch of the aortic arch (a. lusoria) [1%]

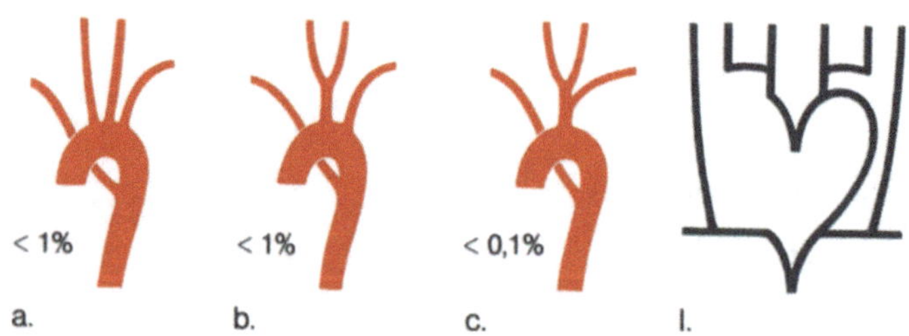

I. development of type 3 a

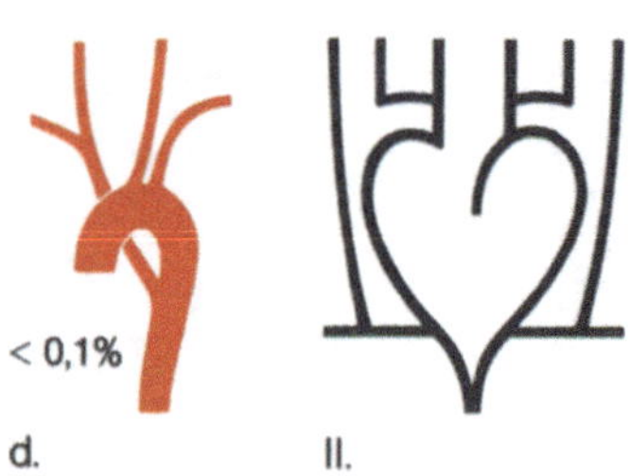

a. otherwise normal
b. combined with a bicarotid trunk
c. combined with a left subclavian artery from the bicarotid trunk

d. right brachiocephalic trunk as the last branch of the aortic arch

II. development of type 3 g

In this situation the right subclavian artery always turns to the right behind the other branches. It is located behind the oesophagus in 80% of all cases, between the oesophagus and trachea in 15% and in front of the trachea or the main bronchi in 5%. This position results in dysphagia or dyspnoea [106, 113, 150*, 191, 230, 238, 247, 249, 415, 442, 468, 506, 605, 611]. In most cases the right recurrent laryngeal nerve is absent, but there are direct branches to the trachea and oesophagus from the vagus. If the right vertebral artery branches from the right common carotid artery (see type 3, left side), a nerve, comparable to the recurrent laryngeal nerve, winds around the artery.

6. Right-sided aortic arch <0.1%

The persistence of the right 4th branchial artery results in this anomaly, which is more frequent in Fallot's tetralogy [54, 61, 73, 110, 243, 248, 264, 306, 357, 564, 568*, 591]. The right-sided aortic arch is the rule in birds and in certain reptiles [234].

a.

a. mirror image of type 1

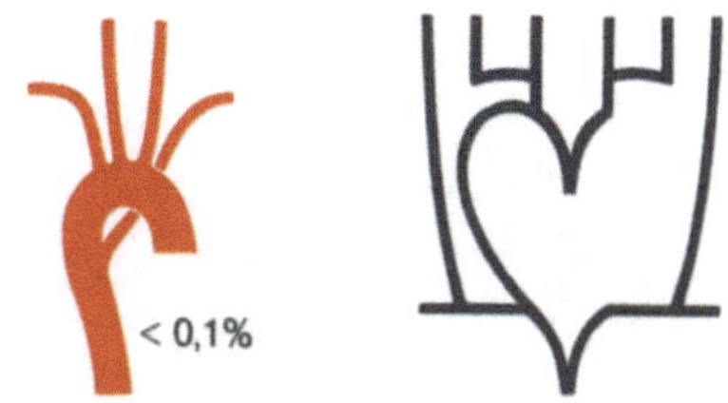

b.

b. mirror image of type 5 a

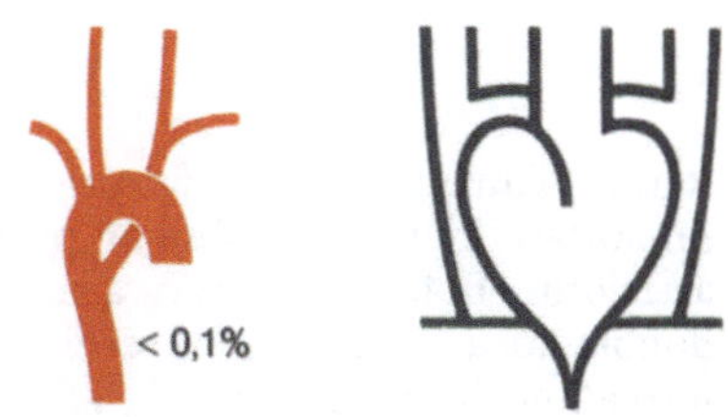

c.

c. mirror image of type 5 d

7. Double aortic arch <0.1%

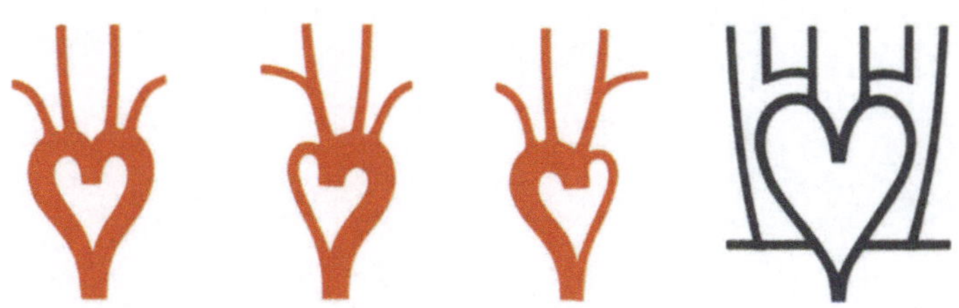

Most of these cases must be classified as malformations, since they have been reported to cause severe functional disturbances of the trachea and oesophagus. However, some rare cases were detected only by chance. Both arches often have different lumens or one aortic arch can be replaced by only a fibrous band. These malformations are often combined with heart defects or a patent ductus arteriosus [39, 61, 67, 154, 155, 192, 193, 194*, 204, 271*, 502, 569, 611, 615, 638]. Double aortic arches are the rule in all branchial breathing vertebrates and also occur in some reptiles [234].

8. Circumflex aortic arch <0.1%

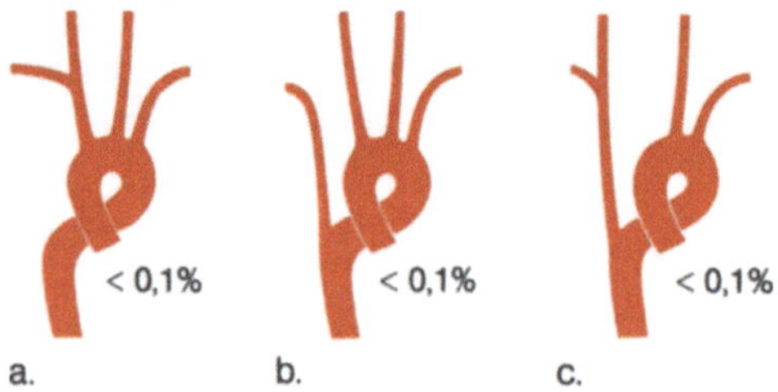

a. left circumflex aortic arch, otherwise type 1
b. left circumflex aortic arch, otherwise type 5a
c. left circumflex aortic arch, otherwise type 5d

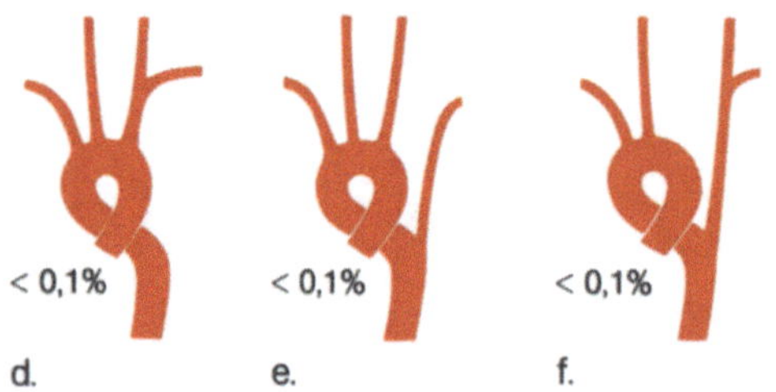

d. right circumflex aortic arch, otherwise type 6a
e. right circumflex aortic arch, otherwise type 6b
f. right circumflex aortic arch, otherwise type 6c

Paramedian merging of the primitive dorsal aortas, the tension of an atypical ductus arteriosus (Botallo's duct) or a rudimentary double aortic arch can all result in such a formation [20, 54, 58, 61, 147, 162, 243, 329, 469, 624]. Obviously these types are more frequent in cases of an arteria lusoria (type 5) or a right-sided aortic arch (type 6) in which the caudal instead of the cranial part of the 4th branchial artery persists. The anomalies in types 8a–f are extremes that have been described, also as regards the loss of the last branch (b, c, e, f).

9. Other variations of the aortic arch <0.1%

a. left subclavian artery originates before the left common carotid artery
b. right subclavian artery originates behind the common carotid artery
c. left external and left internal carotid arteries are separate, direct branches of the aorta
d. left costocervical trunk as a direct branch of the aortic arch
e. left internal thoracic artery as a direct branch of the aortic arch
f. the artery of the thymus is a direct branch of the aortic arch

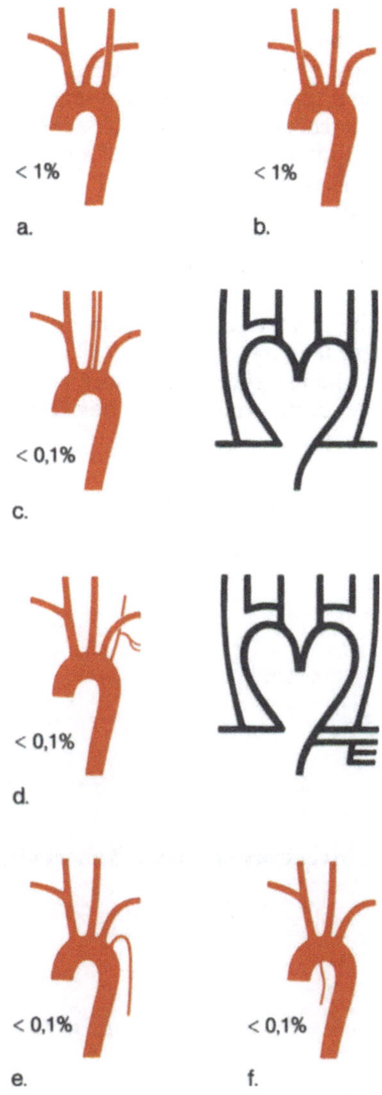

The sequence of branches can vary due to the different growth rates of individual parts of the vessels. The subclavian arteries are pulled downwards by the descending heart, although they are fixed at the upper outlet of the thorax. Many more combinations are possible than are listed. The so-called low division of the carotid artery can be explained by either the first part becoming part of the aortic arch, or the 3rd branchial artery disappearing and the part of the primitive dorsal aorta between the 3rd and 4th branchial arteries persisting. The latter possibility is supported by cases in which the internal carotid artery is totally absent and the common carotid artery alone forms the external carotid artery.
The branches of the costocervical trunk (deep cervical and supreme intercostal arteries) correspond to the primitive segmental arteries (C7, Th1, Th2). Normally they no longer have connections to the aorta and are combined by a longitudinal anastomosis to the subclavian artery.
If the heart descends less caudally than is normal, the so-called cervical aortic arch results [71, 250, 416, 421, 442]. Thus, the topographical position of the aortic arch varies in relation to the vertebral column.
Finally, it must be stressed that only variations, not malformations, have been described. The many malformations, e.g., transposition of the large vessels, the common arterial trunk, the persistent ductus arteriosus, the coarctation of the aorta, were excluded. Some references deal with the so-called interrupted aortic arch only because this can be explained by the same embryological concept [8, 61, 74, 104, 174, 259, 272, 277*, 377, 438, 446, 481, 500, 512, 563, 637, 671, 673].

3 Coronary arteries

In the last two decades the coronary arteries have been studied extensively by coronary angiography, resulting in numerous case reports on different anomalies. For references see recent books, reviews or descriptions of anomalies of specific patterns of general interest [2, 6, 10*, 53, 138, 151, 160*, 187, 205, 219, 220, 229*, 237, 279*, 280*, 418, 424*, 436*, 455*, 470, 544, 584, 618*, 619, 640, 641*, 657, 664].

The anomalies of coronary arteries can be classified by
A. the presence of accessory branches from the aorta and minor deviations in the origins,
B. the presence of only a single ostium of the coronary arteries,
C. the origin from the pulmonary trunk,
D. anastomoses with other arteries and
E. the variability of the posterior interventricular branch.

Many anomalies have no clinical significance under normal circumstances, but they can explain differences in symptoms and ECG findings caused by sclerotic plaques in an anomalous coronary artery or during heart surgery. Anomalous coronary arteries seem to occur more frequently in cases of anomalies of the heart [151, 205]. These are not usually included in the percentages given in the figures.

1. "Normal" situation as described in textbooks ~60%

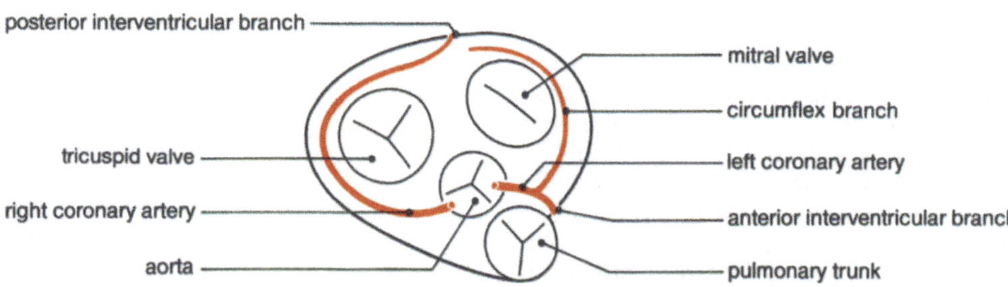

2. Accessory arteries and anomalies of the origin of the aorta ~38%

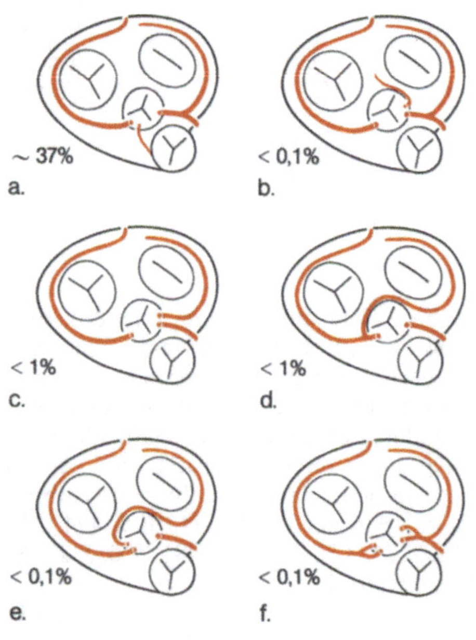

a. artery to the beginning of the pulmonary artery: conus artery
b. additional artery to the septum (a. septi ventriculorum)

c. separate origin of the anterior interventricular branch and circumflex branch of the left coronary artery
d. left circumflex branch but as a branch of the right coronary artery

e. left circumflex branch originates separately on the right side of the aorta
f. two separate origins of both coronary arteries (delta type)

3. Only one coronary artery arising from the aorta [< 1%]

a. left coronary artery branches from the right coronary artery and runs dorsally to the aorta
b. left coronary artery branches from the right coronary artery and runs ventrally to the aorta

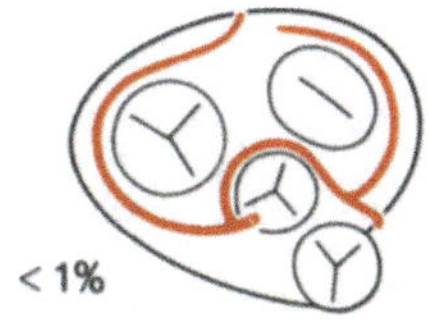

a.

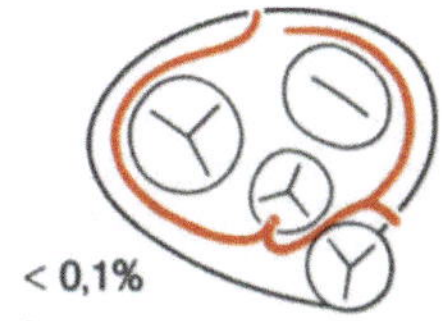

b.

c. right coronary artery branches from the left coronary artery
d. right coronary artery arises from the anterior interventricular branch

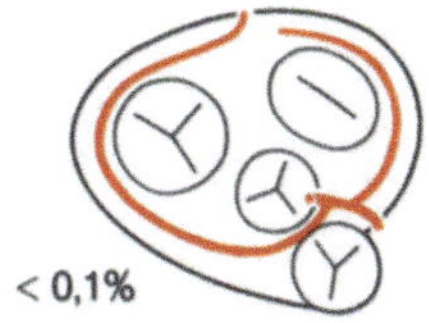

c.

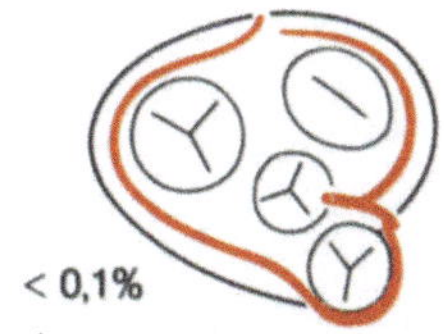

d.

e. left coronary artery is absent, the heart is supplied by an extension of the right coronary artery
f. right coronary artery is absent, the left circumflex branch is extended and supplies the area of the right coronary artery

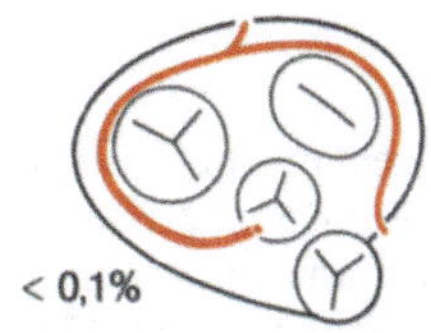

e.

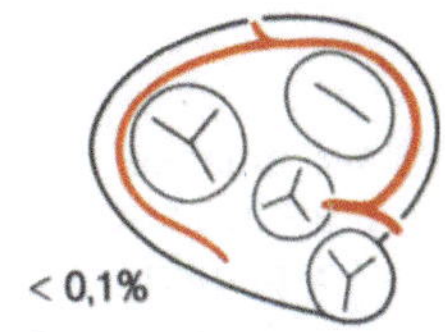

f.

If the aorta has only one coronary orifice, two different types of coronary arteries can be distinguished:
A. Only the first parts of both coronary arteries form a trunk but otherwise all normal branches are present [160*, 298*, 344, 351, 352, 425, 463, 471, 472, 526, 539, 565, 618*, 641*, 647].
B. One coronary artery is totally absent, resulting in an anomalous blood supply to the whole heart by only one coronary artery. Comparable symptoms are found in patients in whom both coronary arteries originate from the same sinus of Valsalva [285, 296, 298*, 305*, 351*, 363, 425, 472, 517*, 565, 577, 618*]. A coronary artery running between the aorta and pulmonary trunk or intramurally can cause intermittent angina [41, 66, 120, 530, 618]. Angiological studies have shown that more patients have only one ostium for one coronary artery than have been reported in anatomical studies. Since coronary angiographies are performed only on patients with angina, a select group of patients are investigated. An anomalous course of a single coronary artery can cause exertional angina, requiring coronary angiography.

4. Origin from the pulmonary trunk 1%

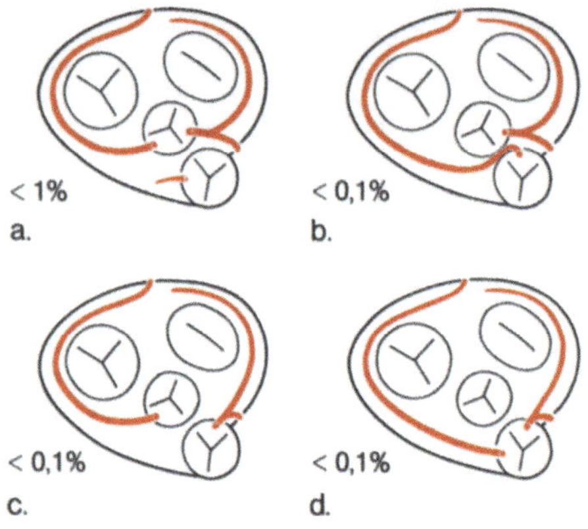

a. accessory branch from the pulmonary trunk
b. right coronary artery from the pulmonary trunk

c. left coronary artery from the pulmonary trunk
d. both coronary arteries from the pulmonary trunk

All six anlagen of valves in the outflow tract can have orifices for coronary arteries, thus explaining the origin of coronary arteries from the pulmonary trunk [232, 589]. The left coronary artery originates from the pulmonary trunk about 10 times more often than the right. This can be explained by the fact that the left orifice is located nearer to the pulmonary trunk [25, 43, 87, 131, 137, 149, 196, 209, 231, 232, 262, 265, 344, 371, 419, 426, 475*, 529, 532, 610, 656, 663*, 674]. The anomalous coronary artery often anastomoses with the normal one and, due to pressure differences, blood is shunted from the normal via the anomalous coronary artery into the pulmonary trunk. Consequently, the afflicted patient does not suffer from low oxygen tension but rather from perfusion pressure in the anomalous coronary artery. In very rare cases coronary arteries can also arise from pulmonary arteries. A few cases have been described in which the coronary artery originated from the brachiocephalic trunk [50, 117, 409] or the bronchial arteries [49, 409, 578, 585].

5. Anomalies of the posterior interventricular branch

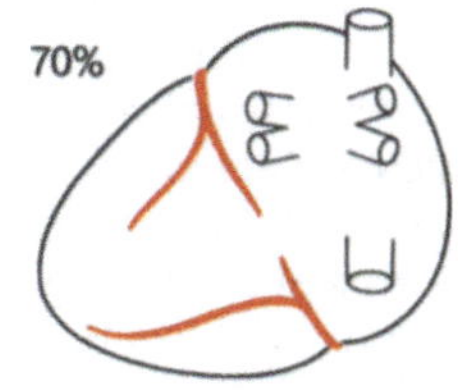

a.

a. “normal” type: the posterior interventricular branch is the terminal branch of the right coronary artery

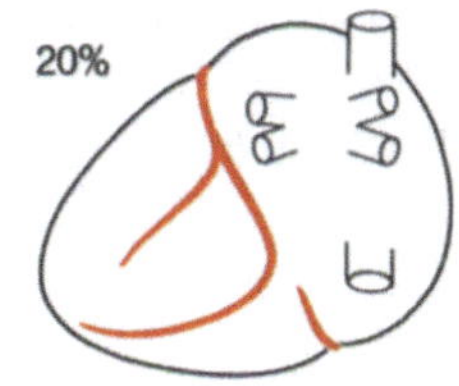

b.

b. “left type” of blood supply: the posterior interventricular branch is the terminal branch of the left circumflex branch

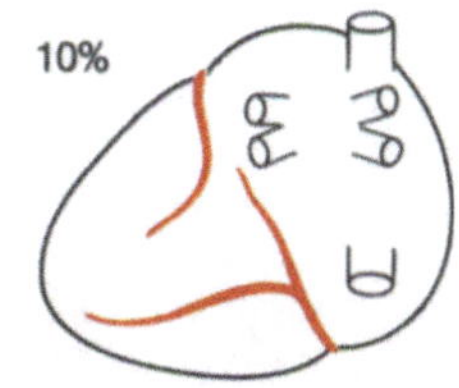

c.

c. “right” type of blood supply: the right coronary artery supplies parts on the left side of the posterior interventricular sulcus

Normally the posterior interventricular artery is a branch of the “normal type” of right coronary artery. If, however, it arises from the left coronary artery, the whole septum is supplied by the left coronary artery (“left type”). The other alternative is the “right type”, in which the right coronary artery reaches the parts of the heart beyond the posterior interventricular sulcus. In many cases the anterior interventricular branch does not end at the apex of the heart, but reaches the posterior parts, resulting in two posterior interventricular branches, one from the cranial and one from the caudal direction [3, 4, 365, 470, 584, 641].

4 Posterior intercostal arteries

The posterior intercostal arteries III to XI, the subcostal artery and the lumbar arteries remain segmental dorsal branches of the descending aorta [70*, 89]. When the heart descends, it causes the more cranial intercostal arteries to take an upward course. Their origins are therefore near to each other, explaining the occurrence of trunk formations of two or more arteries, especially of the 3rd and 4th arteries. The percentages given in Fig. 1c + 1d are the mean of all segments. During embryologic development a longitudinal anastomosis is formed between the segmental arteries C7, Th1 and Th2 and the original connections to the aorta disappear, resulting in the costocervical trunk. As a rule the anastomosis runs ventrally to the ribs, but sometimes it is positioned between the rib and the transverse process, like the vertebral artery which runs through the transverse foramina in the cervical vertebral column [3, 70*, 89, 99, 161, 252, 549].

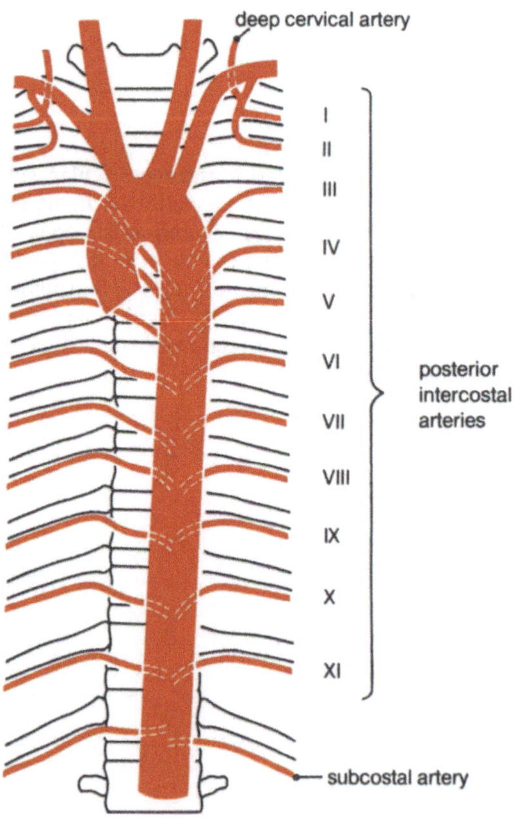

1. Posterior intercostal arteries III–IX and the subcostal artery
(percentages given per segment)

a. separate origins of the intercostal arteries of one segment
b. trunk formation of the intercostal arteries of one segment

a. 83%

b. 2%

c. two intercostal arteries on one side arise from a trunk
d. three or more intercostal arteries on one side arise from a trunk

c. 13%

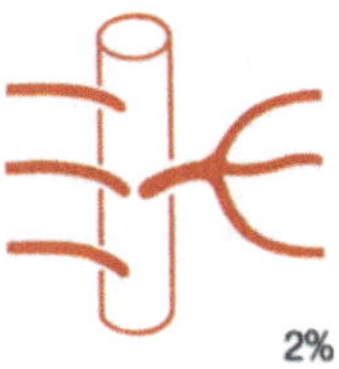
d. 2%

2. The supreme intercostal artery

a. the upper three intercostal arteries (I–III) originate via the supreme intercostal artery from the costocervical trunk of the subclavian artery
b. only the intercostal arteries I and II originate from the supreme intercostal artery

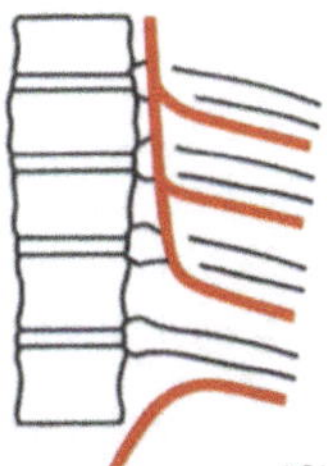
a. 5%

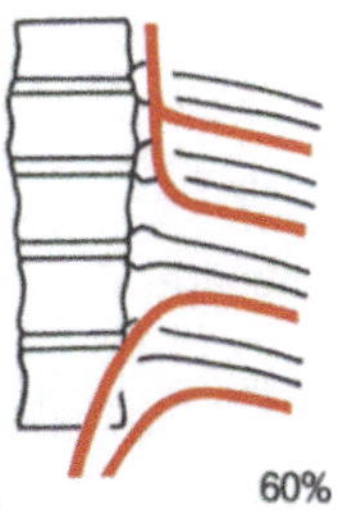
b. 60%

c. only the first intercostal artery is a branch of the costocervical trunk, the second intercostal artery arises directly from the aorta and in about 50% of all cases as a common trunk with the third intercostal artery
d. the posterior intercostal arteries I and II are direct branches of the aorta; sometimes they form a trunk with the 3rd intercostal artery

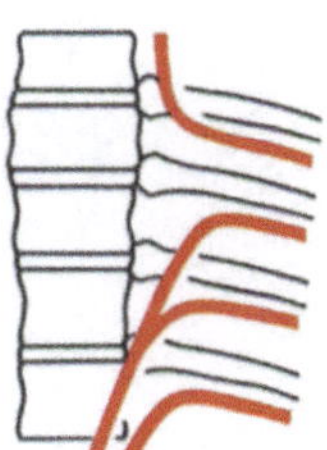
c. 32%

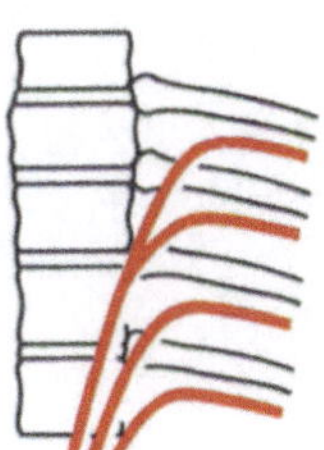
d. 3%

3. The intercostal arteries run dorsally to the ribs ("thoracic vertebral artery")

a. only the supreme intercostal artery lies in a dorsal position
b. dorsal position of the 3rd intercostal artery

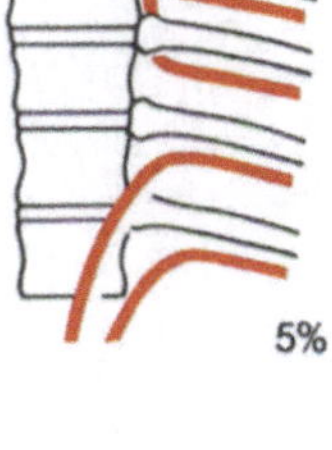
5%
a.

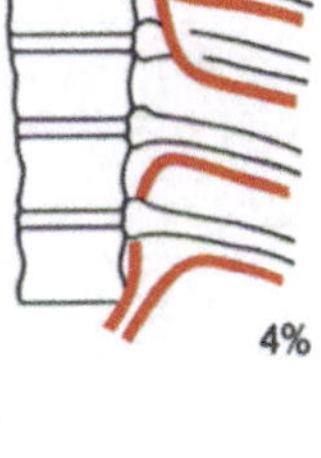
4%
b.

5 Oesophageal arteries

The oesophageal arteries form a longitudinal anastomotic chain in contrast to the radial pattern of arteries in the intestines [15, 82, 119, 197, 562, 601]. The oesophagus can be divided into four segments. The cervical part is usually supplied by the inferior thyroid artery, with more branches from the right than the left side. Other origins of arteries to the cervical part arise from the subclavian artery, the arteria thyroidea ima or the common carotid artery [3, 15, 562*, 601*]. Accessory arteries also come from these arteries. The portion adjacent to the bifurcation of the trachea is supplied by the bronchial arteries, more from the left than the right side. In addition, the 3rd and 4th intercostal arteries give off branches. This results in only three oesophageal branches, contrary to textbooks which often refer to more [15, 82*, 562*, 601*].

1. Basic pattern of blood supply to the oesophagus

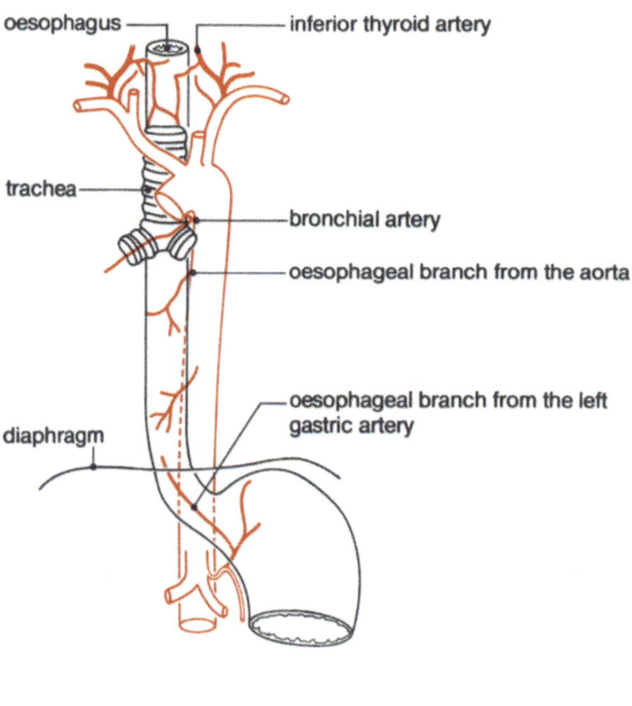

2. Cervical part of the oesophagus

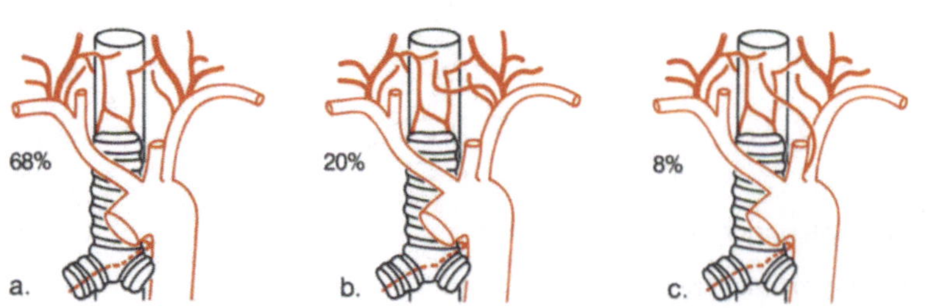

a. branches from the inferior thyroid artery
b. accessory branches from the subclavian artery
c. accessory branches from the carotid artery

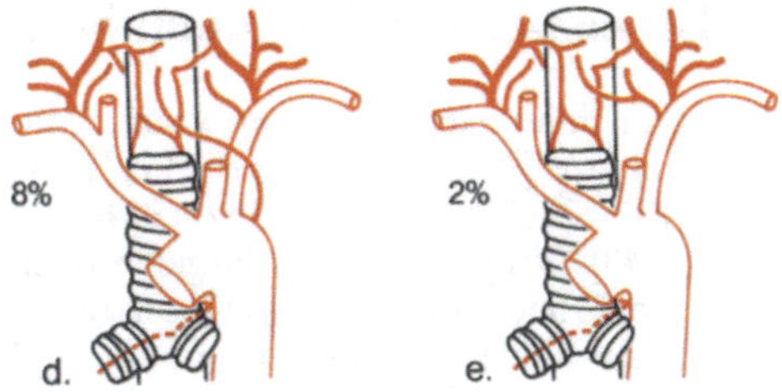

d. accessory branches from the aorta
e. accessory branches from the vertebral artery

3. Thoracic part of the oesophagus

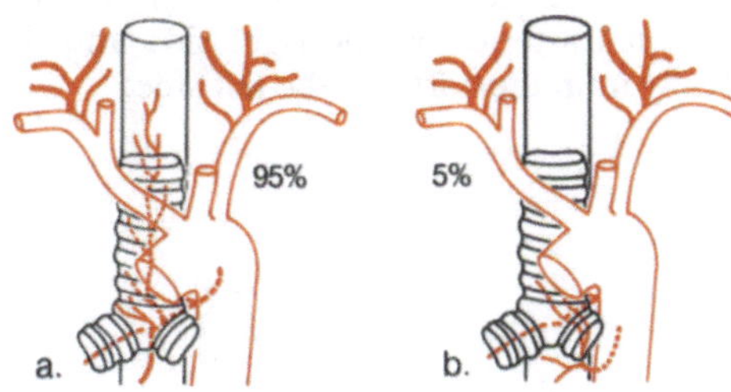

a. branches from the bronchial arteries, two on the left, one on the right
b. direct branches from the aorta as accessory arteries

4. Abdominal part of the oesophagus

The abdominal part of the oesophagus is supplied by 2–3 branches from the left gastric artery. Other origins are the left accessory hepatic arteries and the left inferior phrenic artery [15, 82*, 562*, 601*]. Many oesophageal arteries are relatively short, an important point to consider during surgical mobilization of the oesophagus [399a, 440, 562*].

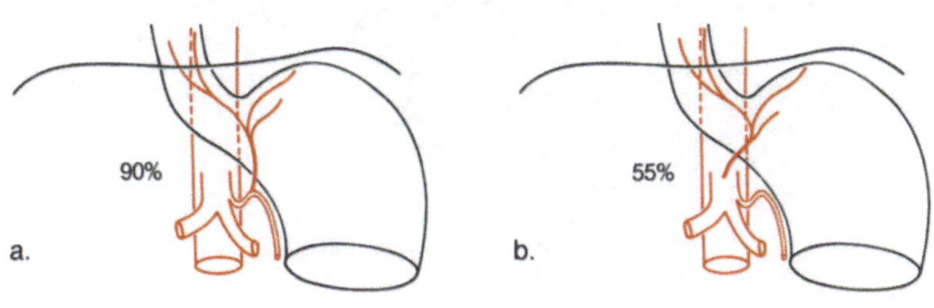

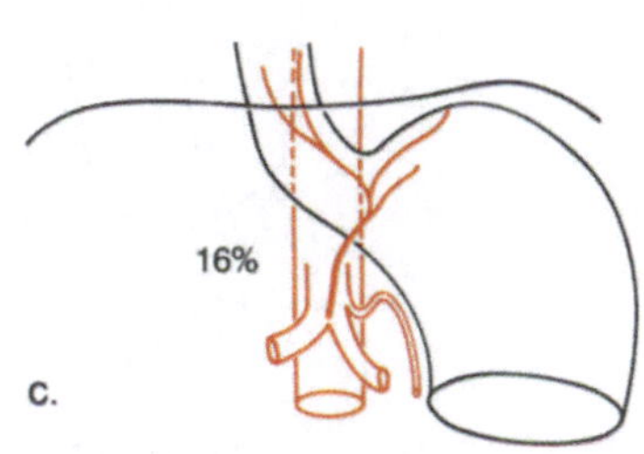

a. major arteries arise from the left gastric artery
b. major arteries arise from the inferior phrenic artery
c. major arteries arise from the celiac trunk

6 Bronchial arteries (Rami bronchiales)

The bronchial arteries vary in origin, number and size [3, 5, 15, 70*, 89*, 111*, 157, 201, 252, 290*, 313, 337, 362, 364, 379, 430*, 543]. They supply the larger bronchi and the bronchopulmonary lymph nodes. There are some anastomoses to the pulmonary arteries [70, 313, 390, 393, 505]. The anomalies fall within three categories.

a. Whereas two bronchial arteries are usually present on the left side, the right side has only one;
b. The bronchial arteries arise from the intercostal arteries on the right side and directly from the aorta on the left side;
c. The right bronchial branches more often run dorsally than ventrally to the oesophagus.

In rare instances, not illustrated here, the bronchial arteries originate from the internal thoracic artery, subclavian artery, supreme intercostal artery, inferior thyroid artery, ascending aorta, abdominal aorta or the inferior phrenic artery [70*, 89, 290, 354, 457]. Coronary to bronchial anastomoses can result in a coronary steal syndrome [410, 578].

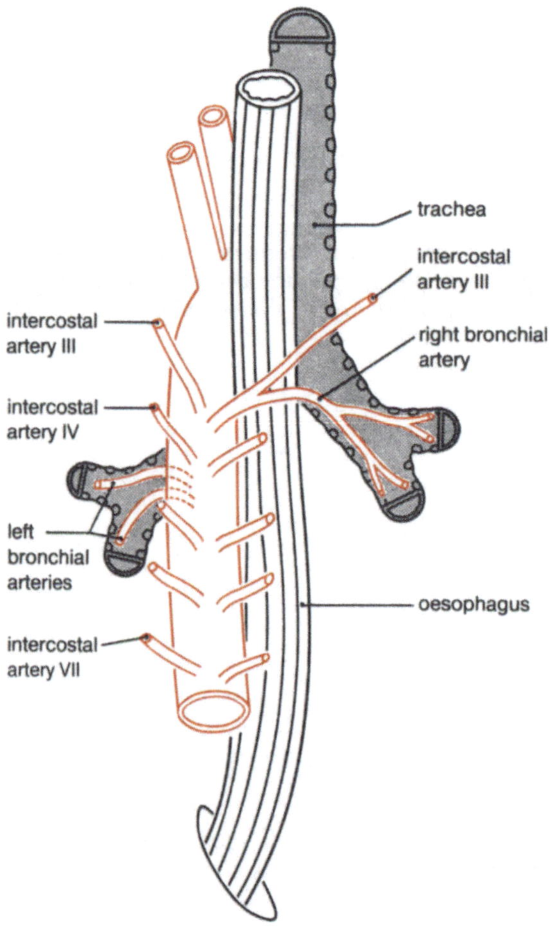

1. Two bronchial branches

a. right bronchial branch from the third posterior intercostal artery (more seldom 4th or 5th); left bronchial branch directly from the aorta
b. both branches directly from the aorta, sometimes with a common trunk

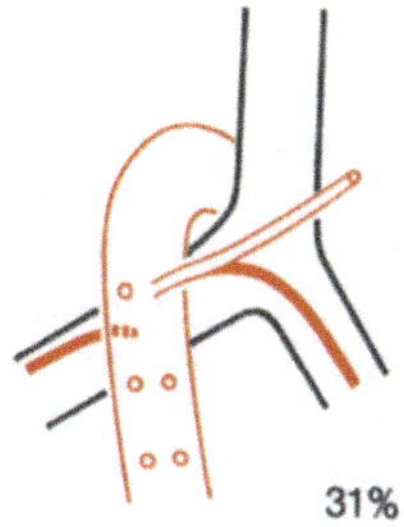

a.

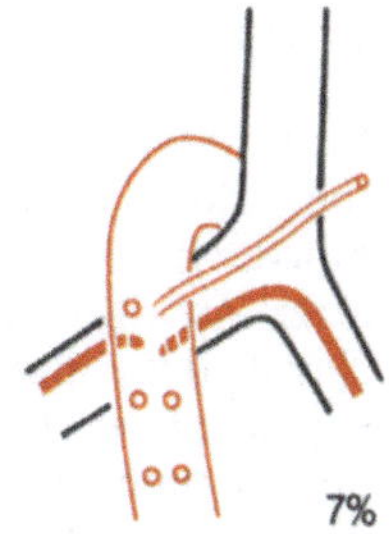

b.

2. Three bronchial branches

a. two left bronchial arteries (mostly from the aorta, less frequently from the right posterior intercostal artery III or IV)
b. two right bronchial arteries

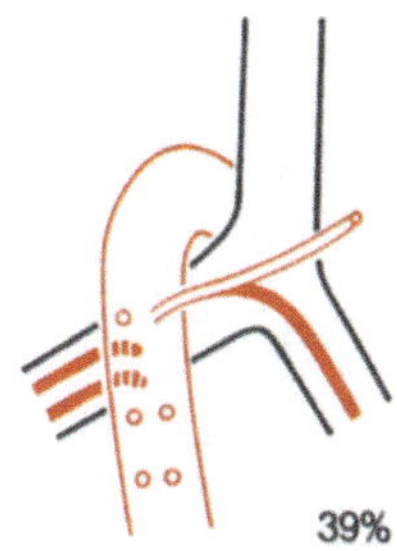

a.

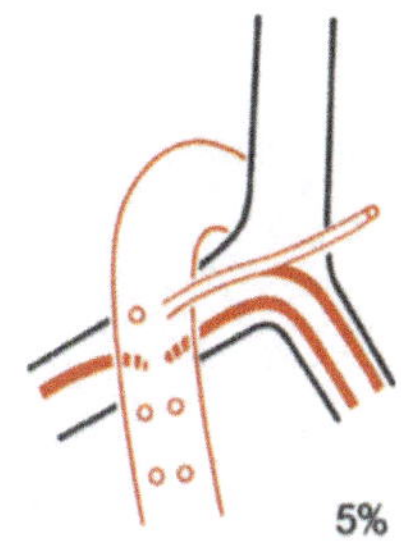

b.

3. Four bronchial branches

a. two right and two left bronchial branches
b. two right bronchial arteries

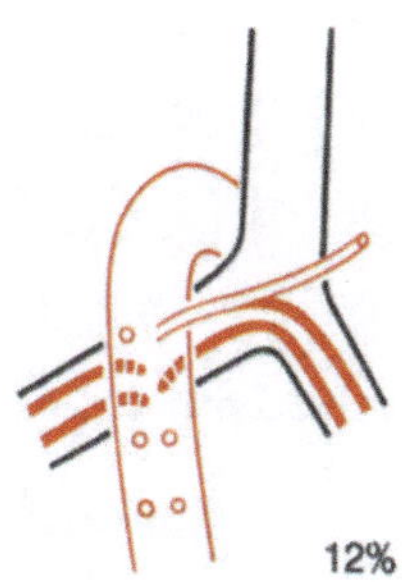

a.

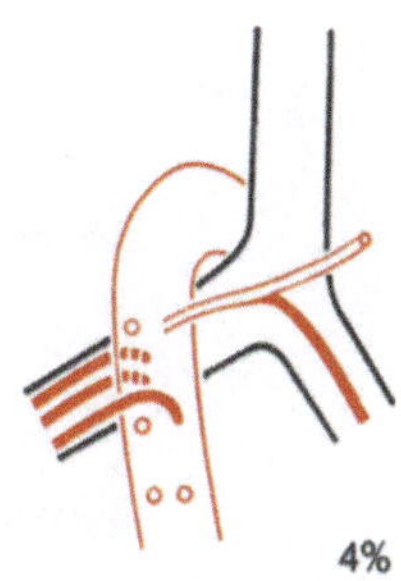

b.

4. Five or more bronchial branches

a. two right and three or more left bronchial branches
b. three right and two or more left bronchial branches

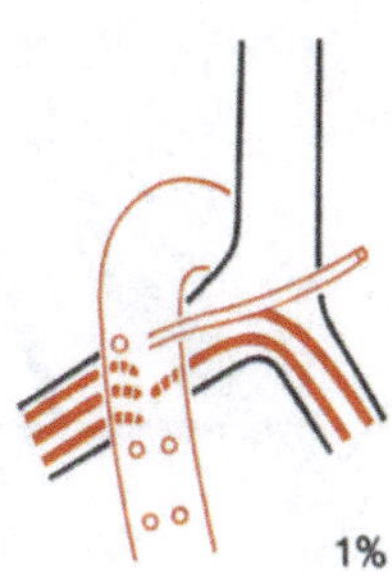

a.

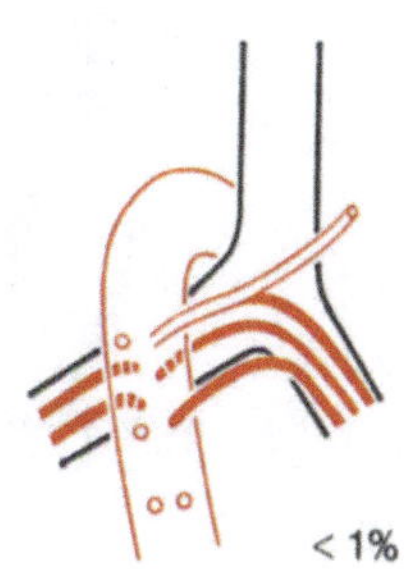

b.

7 Pulmonary arteries

Since anomalies of the pulmonary trunk or the pulmonary arteries are normally combined with heart defects or a patent ductus arteriosus Botalli, they are more aptly classified as malformations. Many can be explained on the basis of embryological development of the aortic arches (p. 3). They can be classified as a systemic arterial blood supply to the lung, an atypical course of a pulmonary artery (sling formation), and the absence of one pulmonary artery.
The pulmonary arteries derive from the 6th aortic arch (p. 3). Whereas its distal part, the connection with the dorsal aorta, disappears, it remains open on the left side until birth as the ductus arteriosus. Many

1. Normal situation and development

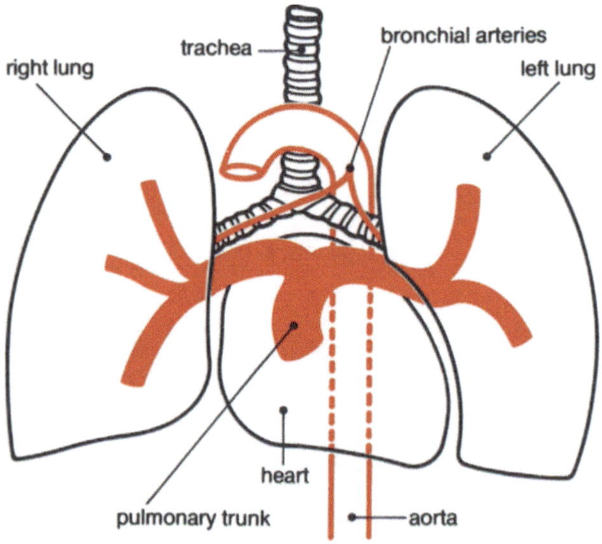

a. normal blood supply to the lung (>99%)

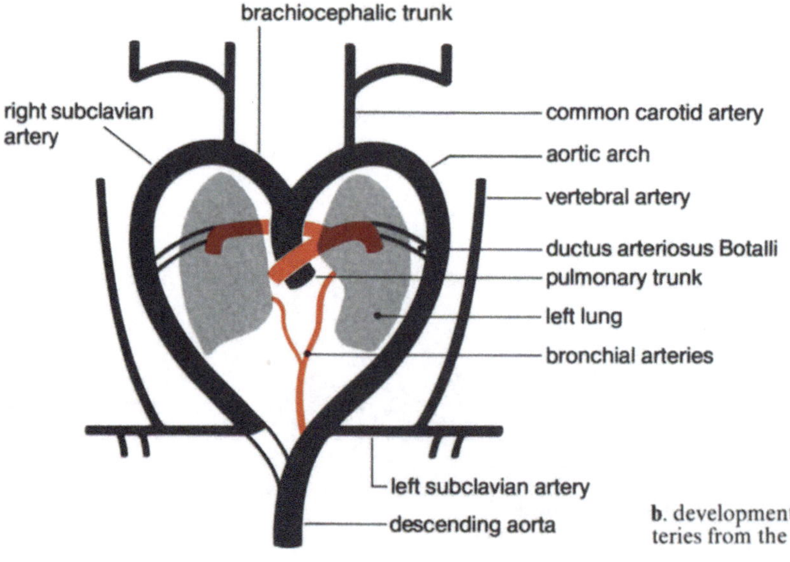

b. development of pulmonary arteries from the 6th aortic arch anomalies can be attributed to the failure of an arch to develop normally or to a precipitous regression or a sudden hiatus of growth [151, 152*, 158*, 206, 478, 496*, 505].

An anomaly in which systemic arteries supply blood to the lung can be explained by the early stage of embryologic development during which branches of the dorsal aortas initially grow into the lung. The distal parts are most often affected by this anomaly, which causes problems of higher systemic blood pressure and flow [158*, 163, 274, 292, 300, 581].

Pulmonary slings are normally caused by an aberrant left pulmonary artery which arises from the right pulmonary artery and passes between the trachea and the oesophagus to the left lung, producing respiratory distress [27, 86, 140*, 217*, 322, 375, 531*].

2. Supply by systemic arteries as branches from the aorta ≪ 0.1%

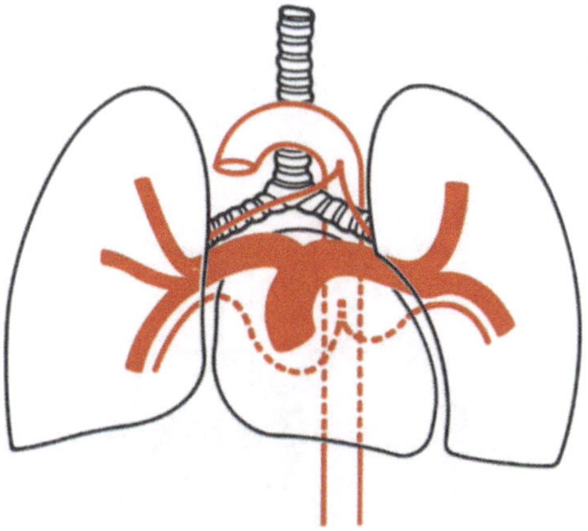

3. Sling formation: aberrant left pulmonary artery from the right pulmonary artery passing between the trachea and the oesophagus to the left ≪ 0.1%

4. Absence of the right pulmonary artery: enlarged bronchial arteries supply the lung ≪ 0.1%

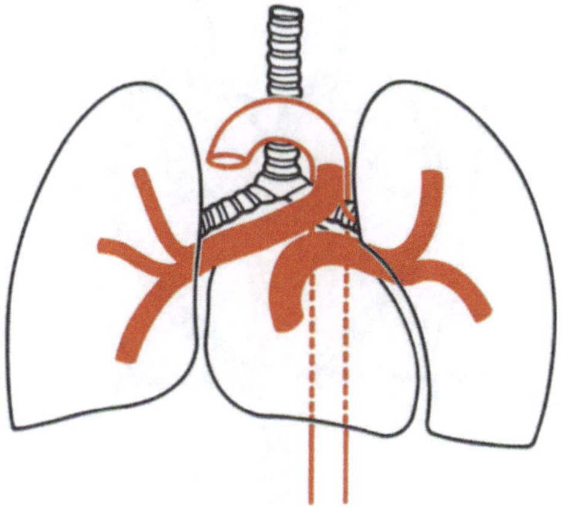

If one pulmonary artery is absent, normally only its proximal part is missing. The more distal parenchymal vessels are generally small but intact [158*, 314*, 478, 496*, 560*]. In these cases either the ductus arteriosus remains patent or a bronchial artery supplies blood to the affected lung.

5. Absence of the left pulmonary artery ≪ 0.1%

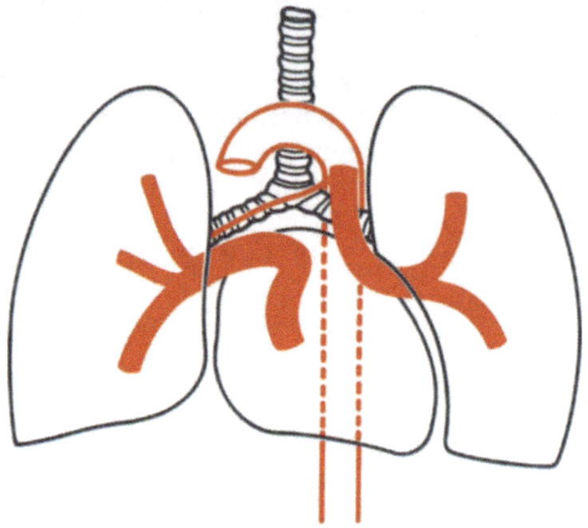

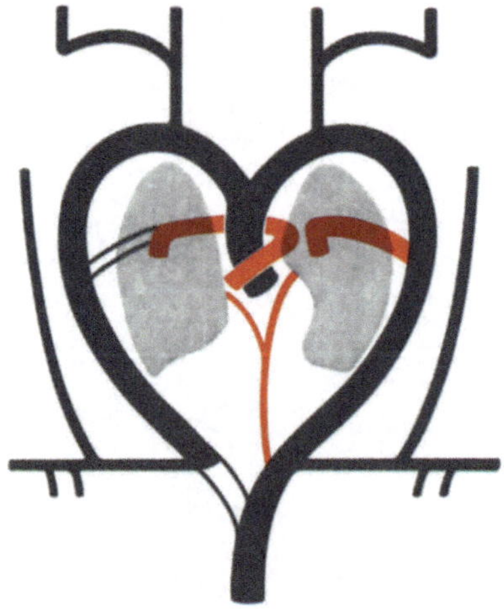

a. the ductus arteriosus remains patent

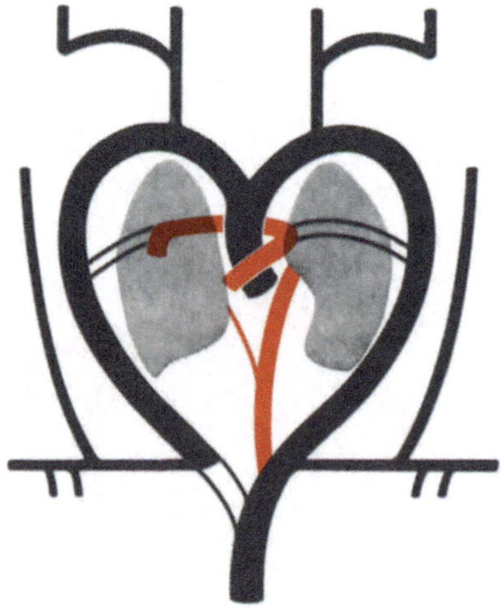

b. enlarged bronchial arteries supply the lung

8 Development of the abdominal aorta

The branches of the abdominal aorta may be divided into three groups: a. dorsal, b. lateral and c. ventral.

a. The dorsal branches are segmental like the branches of the thoracic aorta [3]. The lumbar arteries I–IV remain segmental arteries, but the more important 5th artery forms the main artery for the leg, the common iliac artery. It is comparable to the main vessel supplying the arm, the subclavia, which develops from the 6th cervical segmental artery. Caudal to the 5th lumbar artery, the aorta regresses to the small median sacral artery. The lumbar arteries are not described in detail because their anomalies are comparable to those of the intercostal arteries (pp. 14–15). Whereas the trunk formation of both lumbar arteries from one segment is more common (~4%), the origin of two or more lumbar arteries on one side is much more seldom (~2%) than in the thoracic region [3].

b. The lateral branches supply the kidneys and the genital organs [234, 589]. The mesonephric arteries supply the mesonephros, the adrenal glands and testicles or ovaries. In the second month of pregnancy the adrenal gland is much larger than the kidney and is supplied by three arteries: the superior, middle and inferior suprarenal arteries. One branch runs from the superior suprarenal artery to the diaphragm and another from the inferior suprarenal artery to the kidney. Later the kidneys grow much faster than the adrenal glands; in the neonate they are 1/3 of the kidney weight but in adults only 1/30. Consequently, the suprarenal arteries become less important, and their side branches regress into the main artery. The superior suprarenal artery becomes a small branch of the inferior phrenic artery and the inferior suprarenal artery, a comparably tiny branch of the renal artery. The testicles and ovaries descend, thus elongating their arteries. The origins of these arteries retain their site on the aorta.

c. The ventral branches develop initially as paired vessels. The arteries to the gut merge early to form the celiac trunk and the superior and inferior mesenteric arteries [2, 15, 18, 263, 399a, 403, 495]. The umbilical arteries are also ventral branches which, however, are much larger than all other branches of the aorta in the embryo. The umbilical arteries remain paired vessels, but their origins vary. Longitudinal anastomoses to the segmental artery L5 develop and the initial aorta origins disappear. After birth these arteries are no longer necessary; their distal parts form the lateral umbilical ligaments.

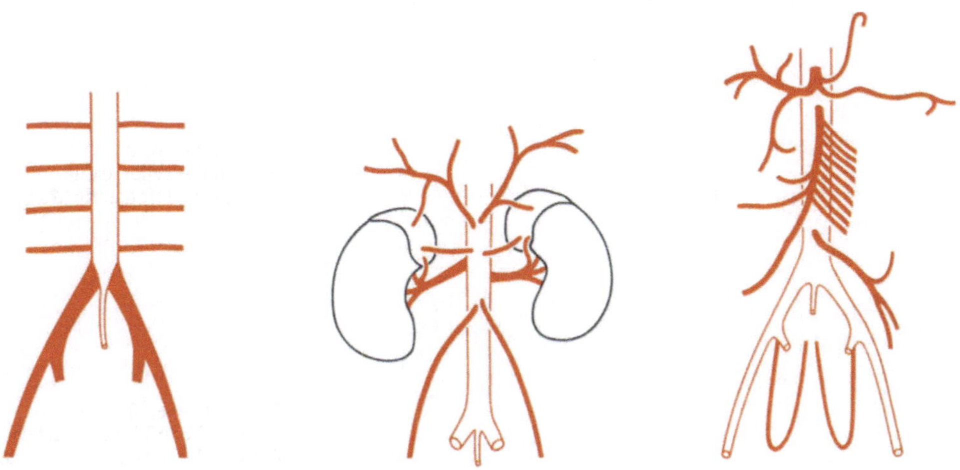

a. b. c.

9 Inferior phrenic arteries

These small arteries supply the diaphragm from the abdominal side. They anastomose with the small superior phrenic arteries, branches of the thoracic aorta, and the musculophrenic and the pericardiophrenic arteries, which are branches of the internal thoracic arteries. The inferior phrenic arteries stem from the mesonephric arteries and thus also supply the adrenal gland via the superior suprarenal artery. This embryologic development elucidates the rare origins of the inferior phrenic arteries from the renal, testicular or ovarian arteries (2 d–f). Their origin from the celiac trunk has been found surprisingly often and can only be explained by a secondary transposition on account of the vessels' different growth rates. The normal pattern described in textbooks accounts for only about a quarter of all cases [3, 15*, 16, 80, 142, 286*, 399*, 399 a, 479, 509, 665].

1. Origin of both inferior phrenic arteries with a common trunk 33%

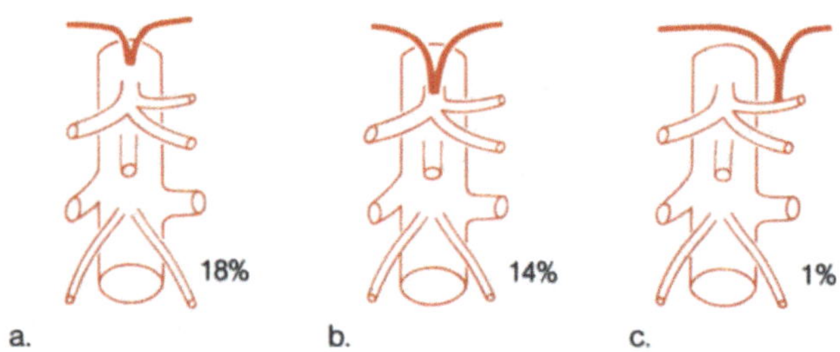

a. from the aorta
b. from the celiac trunk
c. from the left gastric artery

2. Separate origins of the inferior phrenic arteries 67%

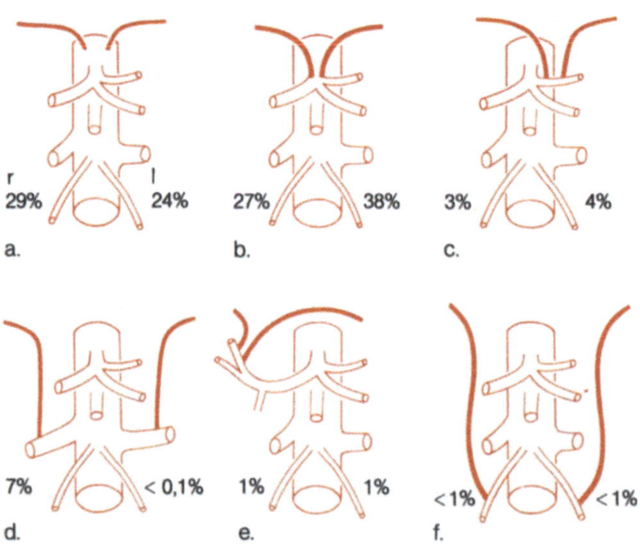

a. from the aorta
b. from the celiac trunk
c. from the left gastric artery

d. from the renal artery
e. from the left hepatic artery
f. from the testicular or ovarian artery

10 Suprarenal arteries

The official anatomical nomenclature cites three arteries supplying the suprarenal gland:
a. the superior suprarenal artery from the inferior phrenic artery,
b. the middle suprarenal artery, a direct branch of the aorta, and
c. the inferior suprarenal artery, a branch from the renal artery.
However, careful preparations after dye injection or corrosion cast preparations give a much more complex picture. The suprarenal arteries are always multiple, and they branch before entering the gland. They could be compared with spokes of a wheel, the suprarenal gland serving as the hub. Up to 60 arterial branches have been counted to enter the gland [3, 16, 142*, 228, 286, 381, 398, 399*, 479]. The venous return, however, takes place via only one suprarenal vein; accessory veins are rarely found.
The inferior suprarenal arteries are in part derived directly from the renal arteries and in part from the renal capsule, which is not only supplied by the renal artery but also by branches from the testicular, ovarian or lumbar arteries. Even a direct origin of the inferior suprarenal artery from the testicular and ovarian arteries has been described [3, 16, 142, 399].

1. Blood supply from three different origins 34%

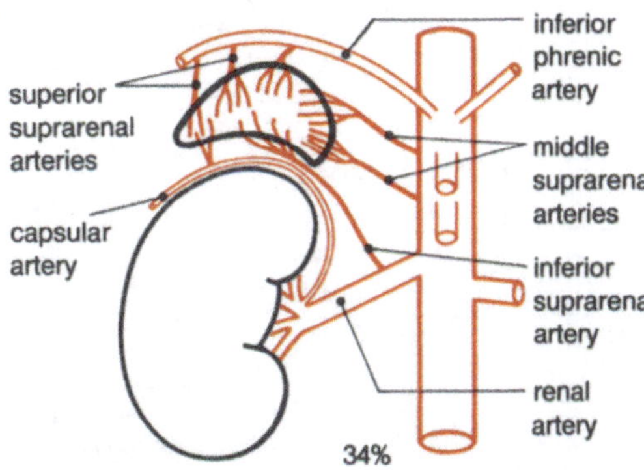

2. Blood supply from two different origins 61%

a. from the inferior phrenic artery and the aorta (more often left than right)
b. from the inferior phrenic artery and the renal artery (more often right than left)
c. from the aorta and the renal artery

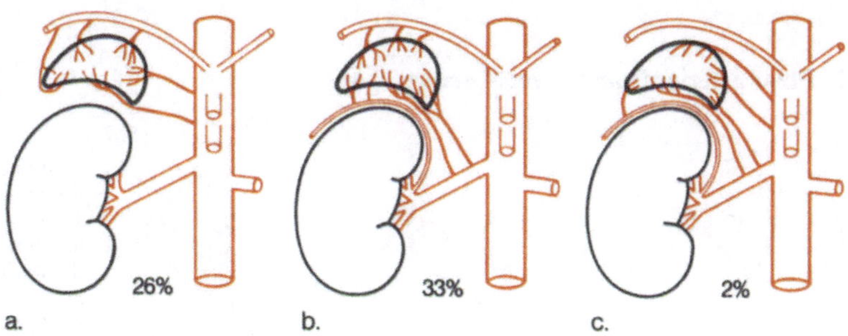

3. Blood supply from one vessel only 5%

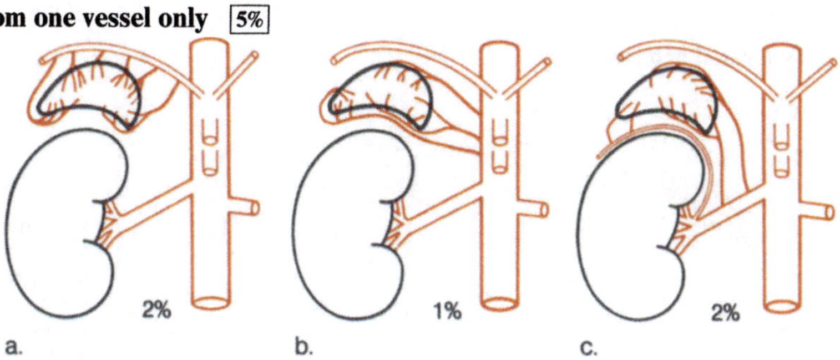

a. from the inferior phrenic artery
b. from the aorta
c. from the renal artery

11 Renal artery

The renal artery normally divides into two main trunks, which after branching in most cases into 5–7 arteries (range 2–10) enter the parenchyma of the kidneys [e.g. 3, 29, 142*, 399*]. They have no anastomoses within the kidneys and are "end arteries", which supply individual segments or parts of segments. Polar arteries are therefore not auxiliary, supplementary or accessory arteries as they have been called, for they replace the normal segmental branch. Polar arteries and multiple renal arteries are important in kidney transplantation [180, 396, 535, 669].

Although renal segments have been included in the official anatomical nomenclature, their functional relevance is still debatable since the segments are based only on the arterial supply [165, 211, 236, 360, 494*].

1. "Normal" pattern as described in textbooks 59%

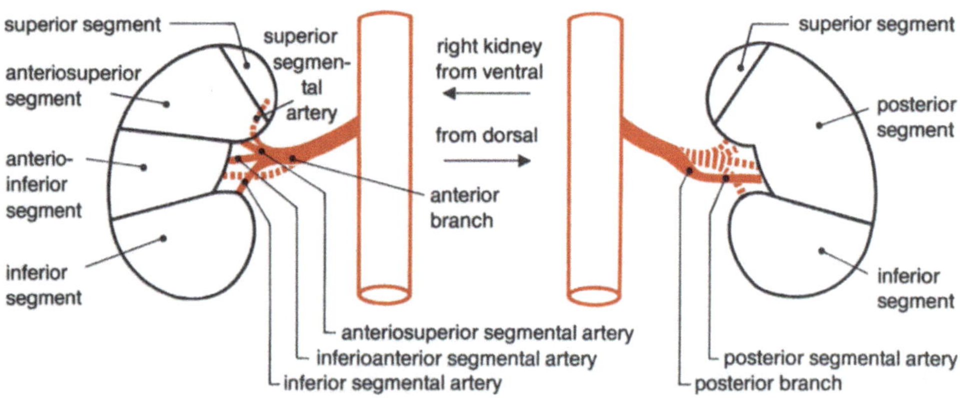

2. Polar arteries from the renal artery 15%

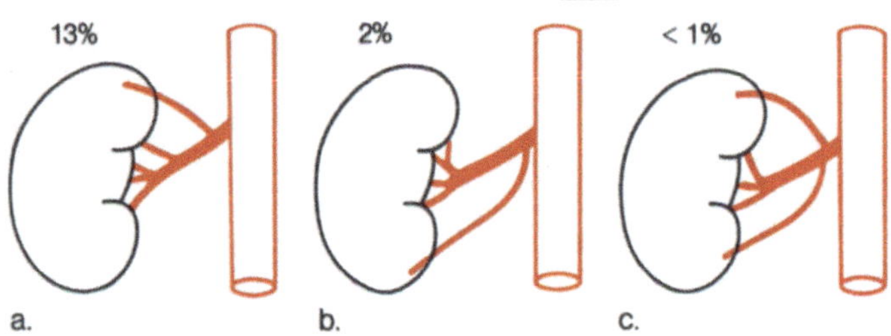

a. upper polar artery from the main stem of the renal artery
b. lower polar artery from the main stem of the renal artery
c. upper and lower polar arteries from the main stem of the renal artery

3. Two renal arteries 22%

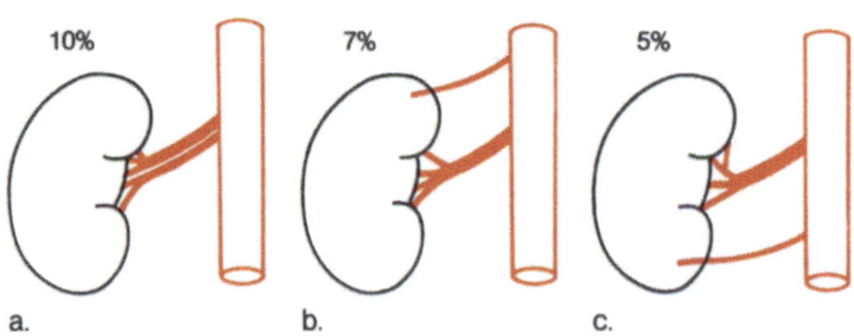

a. both renal arteries run to the hilus of the kidney
b. upper polar artery from the aorta
c. lower polar artery from the aorta

Atypical origins of the renal arteries are found more often in cases of ectopic or horseshoe-shaped kidneys than in kidneys in normal positions [30, 404].
Ectopic origins of renal arteries have been described in individual cases from the following arteries: median sacral, external and internal iliac, lumbar, superior and inferior mesenteric, branches of the celiac trunk and even cranial to the celiac trunk from the aorta [23, 28, 195, 212*, 538, 602, 614].
The high frequency of polar arteries, multiple arteries and the origin of renal arteries from the aorta at different heights or from other arteries can be explained by the embryological development of the kidney. Many segmental arteries supply the pronephros, followed by the more cephalad mesonephros and finally the metanephros. There is a continuous involution of arteries and growth of new renal arteries. The persistence of some of these segmental arteries results in multiple renal arteries, polar arteries or an abnormal origin of a renal artery [3, 15, 16, 17, 30, 63*, 116, 170, 195, 228, 235, 245, 273, 289*, 399*, 402, 408, 493*, 589, 658].

Although the right kidney has a more caudal position than the left kidney, the right renal artery originates in general more cranial to the aorta than the left renal artery [e.g. 3, 16, 17, 289, 658]. The right renal artery is longer than the left one because of the position of the aorta on the left of the vertebral column. The right renal artery usually runs dorsally to the inferior vena cava and only ventrally in about 4% of all cases when a single renal artery is present. In patients with two right renal arteries (in about 30%) one artery crosses the vena cava ventrally.
There is considerable disagreement in the literature on the frequency of multiple renal arteries. This can in part be explained by the technical differences, e.g. anatomical or corrosion cast preparation, aortographies or selective renal arteriograms. In some patients with hypertension or hydronephrosis a higher frequency of multiple renal arteries was described [59, 115, 121, 199, 338, 520]. Multiple arteries sometimes occurred more frequently on the left side. The percentages given in the figures are the mean for both kidneys.

4. Three or more renal arteries 4%

a. all three renal arteries run to the hilus
b. upper polar and two hilar arteries from the aorta
c. lower polar and two hilar arteries from the aorta

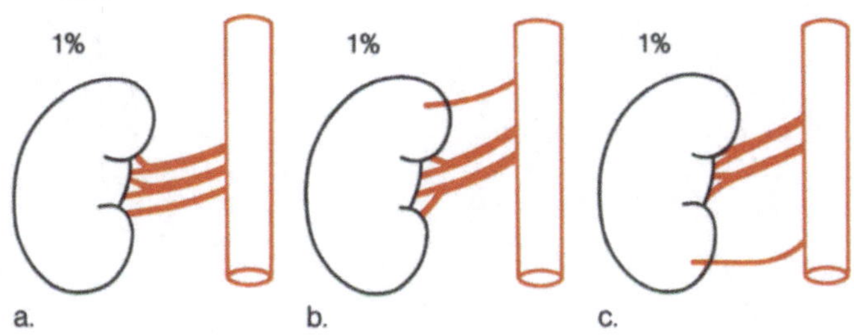

d. upper and lower polar artery from the aorta
e. four renal arteries (because of the rare incidence no further classification is given)
f. five renal arteries (numerous combinations)

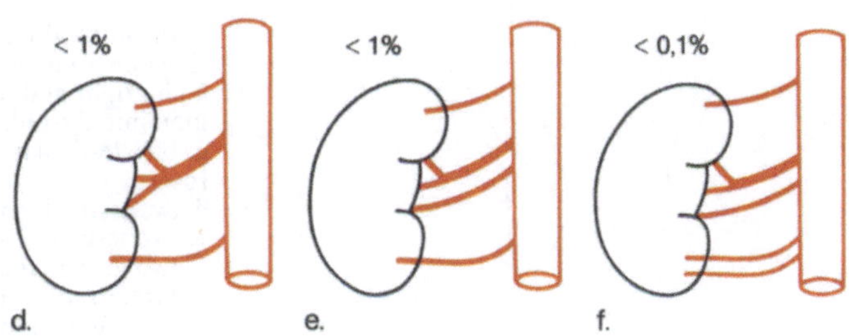

12 Testicular artery

In older studies and textbooks the testicular and ovarian arteries were subsumed under the term internal spermatic artery [3, 15, 448]. Thus, in some studies on variations of these arteries the sex was not mentioned in which the abnormalities were observed [3, 108, 399, 501, 565]. In general, deviations in the origin and course of the testicular and ovarian arteries seem to be comparable. Only the testicular artery is shown in the figures, since there are more studies on anomalies of the testicular arteries [153, 246, 282, 316, 467, 667]. Only a small number of groups have done routine angiographies of the small testicular arteries [68, 445]. In some rare cases the testicular arteries arch over the renal arteries first before they turn towards the inguinal canal [3, 434, 447]. An anastomosis between the artery of the ductus deferens and the testicular artery has sometimes been found along the epididymis.

1. Testicular arteries originating only from the aorta [83%]

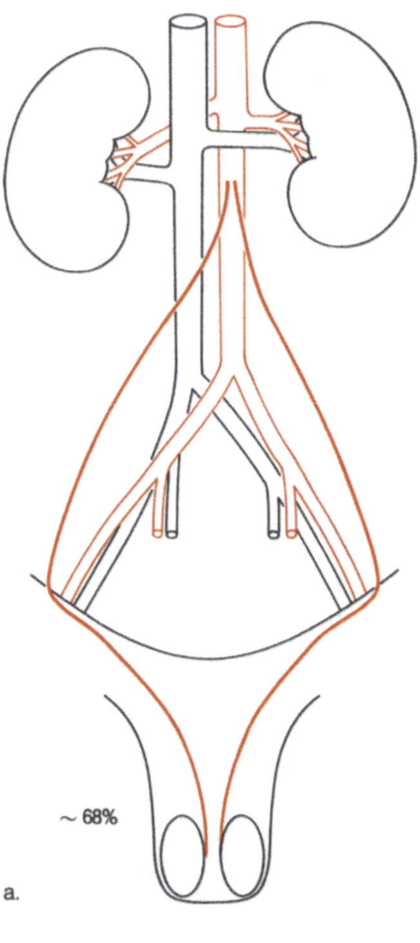

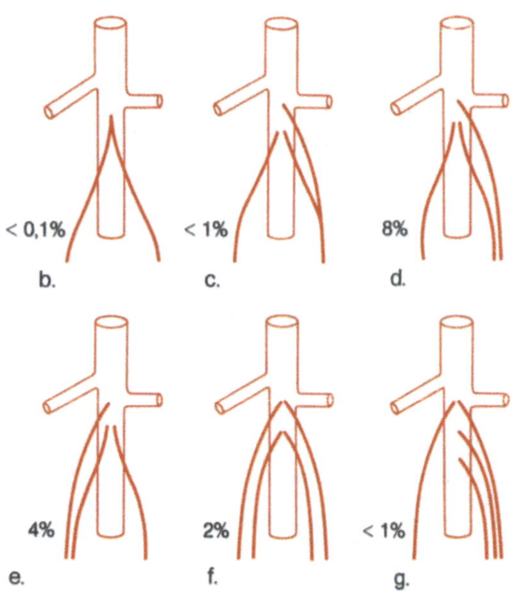

a. the normal figure found in textbooks; on each side one testicular artery arises from the aorta
b. the right and left testicular artery form a common initial trunk
c. one testicular artery has two or more joining roots
d. two testicular arteries on the left side
e. two testicular arteries on the right side
f. two testicular arteries on each side
g. three testicular arteries are found on one side (more often on the left than on the right side)

2. Testicular arteries also originating from the renal artery 17%

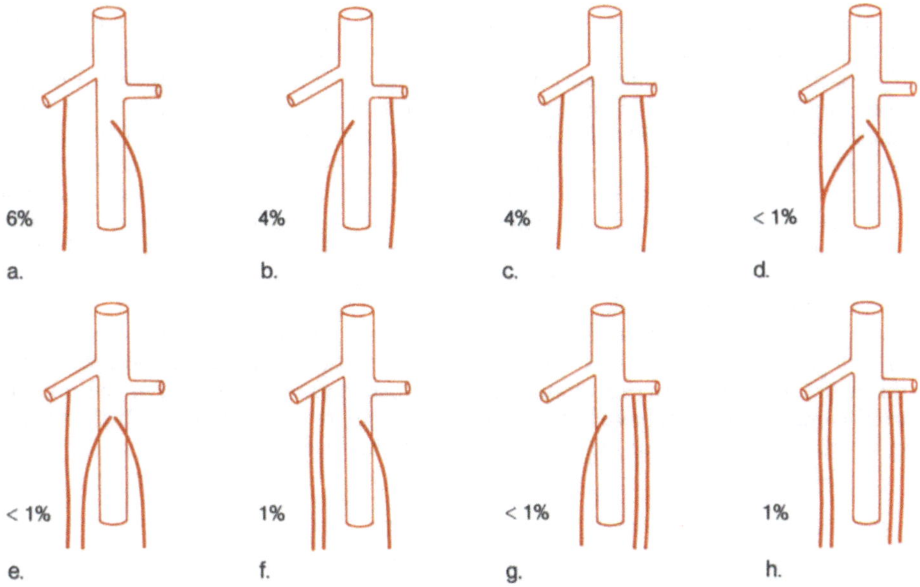

a. right testicular artery from the right renal artery
b. left testicular artery from the left renal artery
c. both testicular arteries from the renal arteries
d. one artery stems from both the aorta and a renal artery
e. two testicular arteries on one side, one from the aorta, the other from the renal artery
f. two right testicular arteries from the right renal artery
g. two left testicular arteries from the left renal artery
h. two testicular arteries on both sides from the renal arteries

3. Testicular arteries from other vessels < 1%

Several cases of testicular arteries arising from the middle suprarenal artery have been described, but origins from lumbar arteries, the common iliac, internal iliac or inferior epigastric arteries were found in only individual cases.

4. Course of the testicular arteries

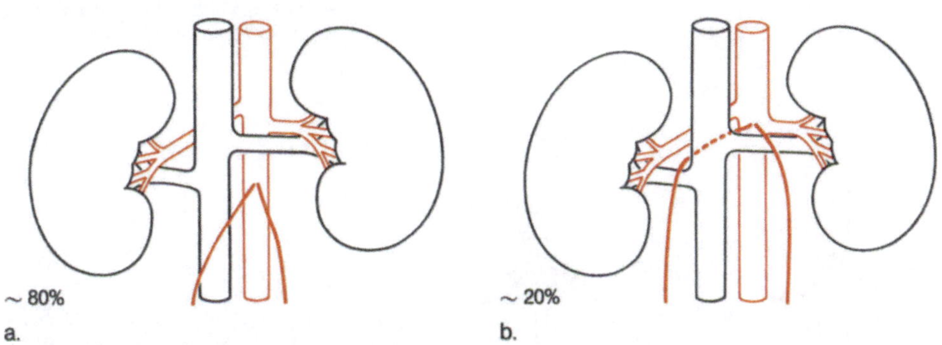

a. the testicular arteries arise caudally to the renal veins
b. the testicular arteries run ventrally to the renal veins (especially when a high origin is present or multiple arteries occur). The right testicular artery is located dorsal to the inferior vena cava in such cases. These anomalies can impede the venous return [3, 316, 434, 447, 458].

13 Celiac trunk

Of the three main branches of the celiac trunk, the common hepatic artery has the highest frequency of anomalies. For details of anomalies of the hepatic, left gastric and splenic artery see pp. 32–33, 36–37, 39. When a complete trunk is described (type 1), this does not mean that the common hepatic artery has all "normal" branches, but only that a major part of the common hepatic artery, e.g. the gastroduodenal artery, arises from the celiac trunk. An accessory hepatic artery from the superior mesenteric or left gastric artery was not included in this figure. As in type 3a (hepatomesenteric trunk), only those cases have been included in which no major part of the common hepatic artery comes from the celiac trunk. A clear-cut differentiation between type 1a and 1b is not always possible. The most frequent accessory branch is the inferior phrenic artery (in about 50%), see p. 24. In this figure only variations of the typical branches of the celiac trunk are included [2, 3, 15, 36, 65, 91, 167, 184, 218*, 263, 367*, 383*, 395, 399a*, 435a, 450, 456, 495*, 515, 528*, 628, 645, 655, 676].
The origins from the aorta of the celiac trunk and the superior mesenteric artery are close to each other and they quite often form common trunks. For the sake of clarity the origins are drawn further apart than is the case in situ.

1. Complete celiac trunk (gastrohepatolienal trunk) 84%

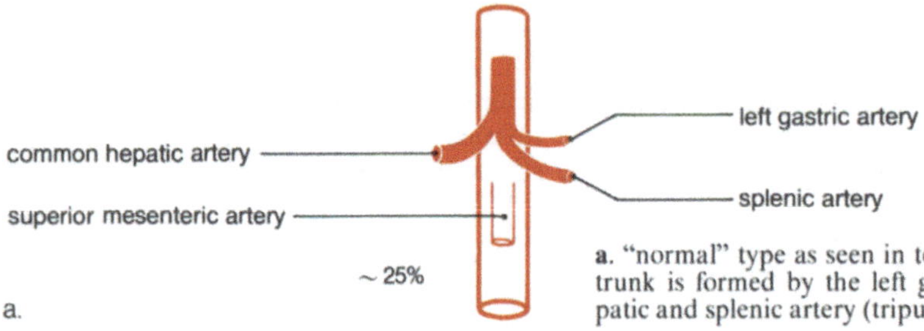

a.

a. "normal" type as seen in textbooks: the celiac trunk is formed by the left gastric common hepatic and splenic artery (tripus Halleri)

b.

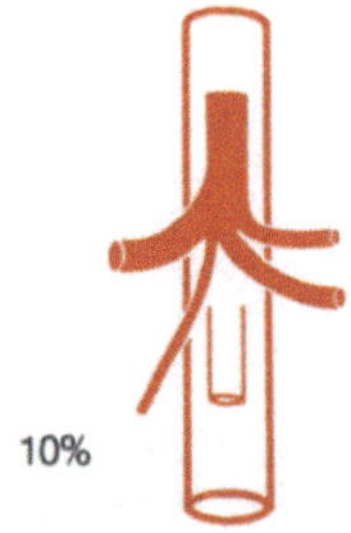

c.

b. the celiac trunk leaves the left gastric artery as a side branch and divides more distally into the common hepatic and the splenic artery
c. the celiac trunk has four main branches (tetrapus), the fourth being an additional dorsal pancreatic artery

2. Incomplete celiac trunk 9%

a. hepatosplenic trunk; the left gastric artery is a separate branch of the aorta
b. gastrosplenic trunk; the common hepatic artery is absent or an independent branch of the aorta

a.

b.

c. gastrohepatic trunk; the splenic artery is an independent branch of the aorta
d. no trunk formation

c.

d.

3. Common origin of main branches of the celiac trunk with the superior mesenteric artery 7%

a. gastrohepatosplenomesenteric trunk
b. hepatosplenomesenteric trunk

a.

b.

c. hepatomesenteric trunk
d. splenomesenteric trunk

c.

d.

14 Hepatic arteries

Like the lung, the liver can be subdivided into lobes and segments which are supplied by end arteries. These arteries, however, frequently have accessory branches. The branching pattern of the common hepatic artery into a left and right hepatic artery,

1. Arterial blood supply of the liver from the celiac trunk only 76%

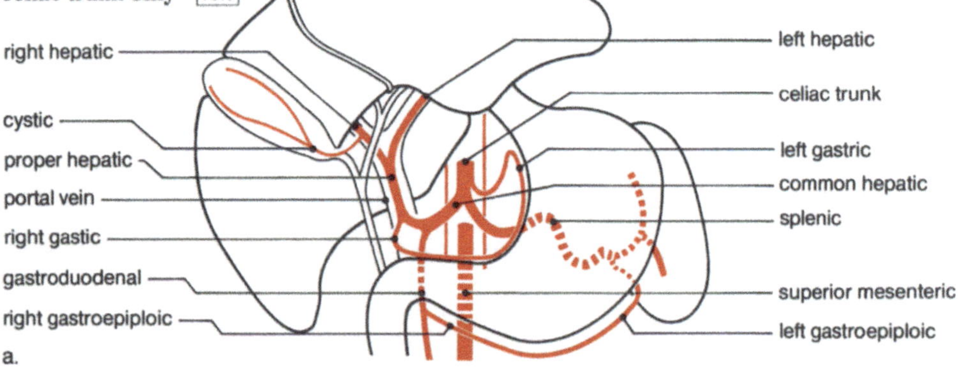

a.

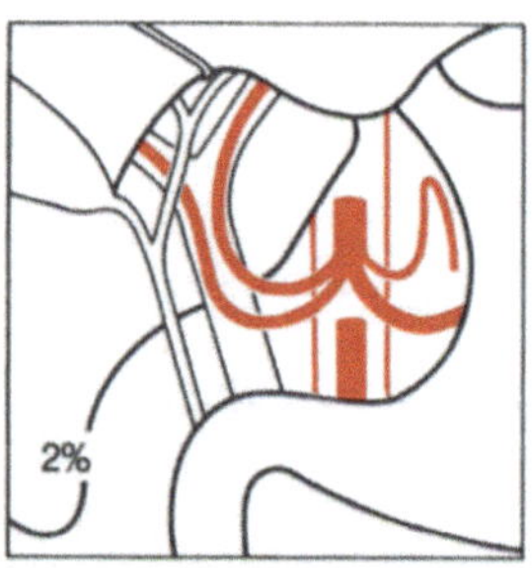

b.

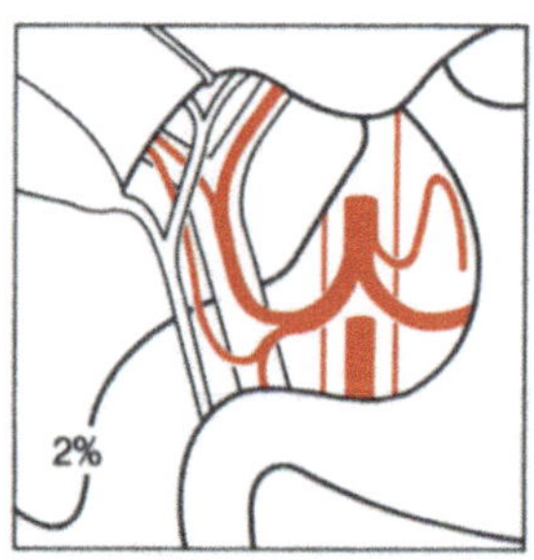

c.

a. "normal" type shown in textbooks: the common hepatic artery is a branch of the celiac trunk or originates direct from the aorta

b. accessory hepatic artery from the celiac trunk or direct from the aorta or a separate origin of the right and left hepatic arteries
c. accessory hepatic artery from the gastroduodenal artery

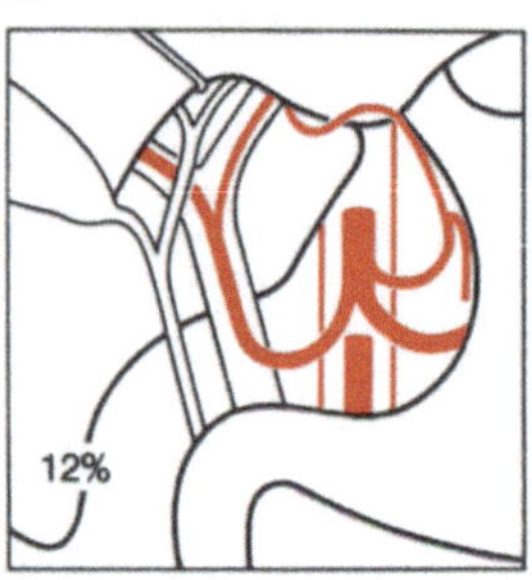

d.

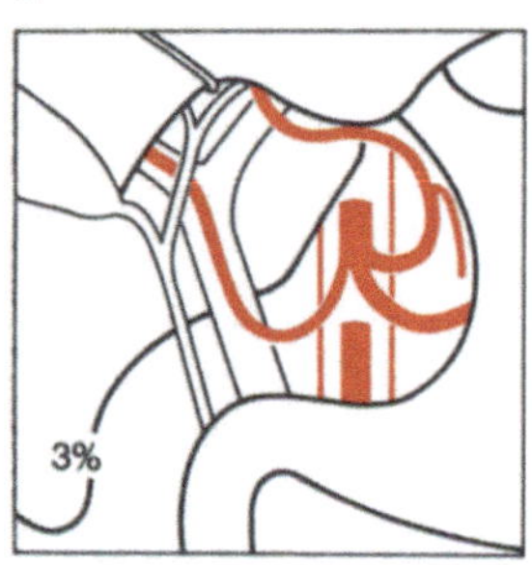

e.

d. accessory left hepatic artery from the left gastric artery
e. replaced left hepatic artery from the left gastric artery

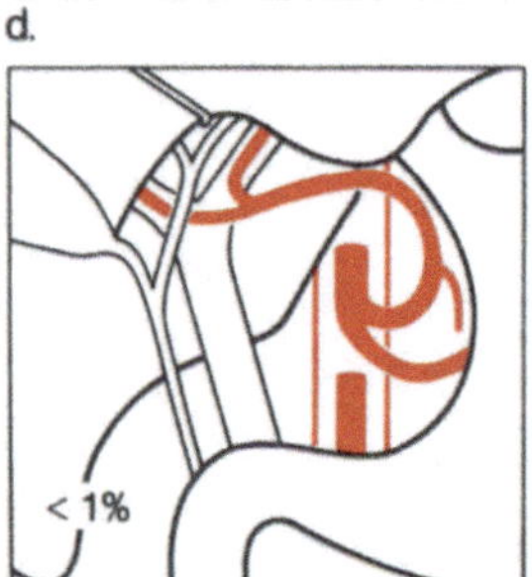

f.

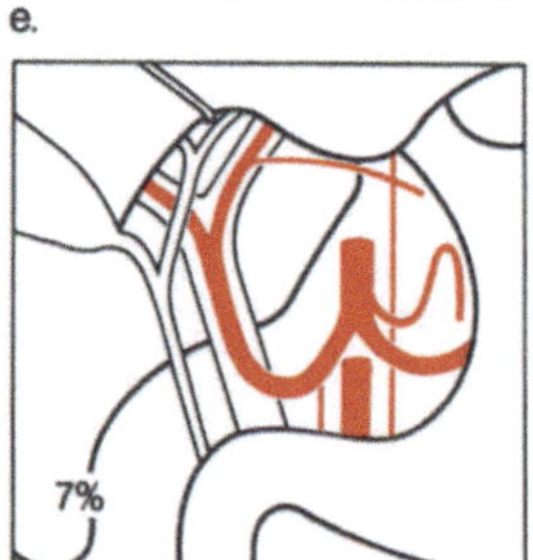

g.

f. common hepatic artery from the left gastric artery
g. accessory left gastric artery from the left hepatic artery

corresponds to the left and right lobe. The border between the functionally defined lobes is not the anatomical border but the line from the gallbladder to the sulcus of the inferior vena cava. Thus, the caudate and quadrate lobe are part of the left lobe and supplied with arterial blood in 90% of all cases by the left hepatic artery. The left lobe can be divided further into a medial and lateral segment and the right lobe into an anterior and posterior segment [2, 3, 15, 18, 102, 105, 167, 175, 177*, 208, 218, 223, 256, 263, 302, 321, 367*, 378, 394, 395, 399 a*, 400*, 401*, 435 a, 449, 450, 452, 456, 491, 495*, 507, 528, 546, 600, 630*, 665, 676].

2. The superior mesenteric artery also supply the liver 24%

a. accessory right hepatic artery from the superior mesenteric artery
b. replaced right hepatic artery from the superior mesenteric artery

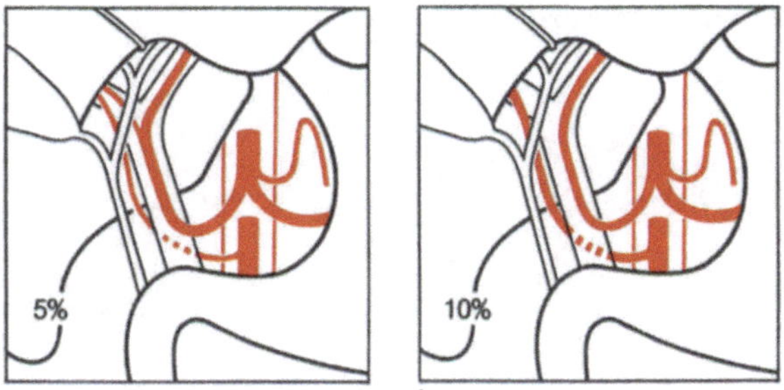

a. b.

c. common hepatic artery from the superior mesenteric artery (hepatomesenteric trunk)
d. common hepatic artery from four-branch trunk (gastrohepatosplenomesenteric trunk) or an hepatosplenomesenteric trunk

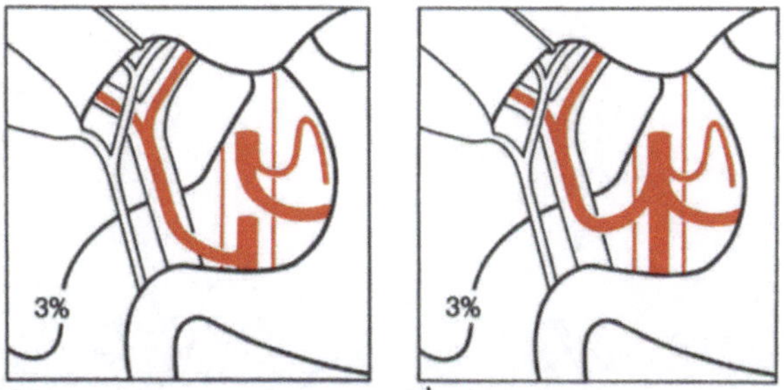

c. d.

e. accessory right hepatic artery from the superior mesenteric and accessory left hepatic artery from the left gastric artery
f. accessory right hepatic artery from the superior mesenteric and replaced left hepatic artery from the left gastric artery

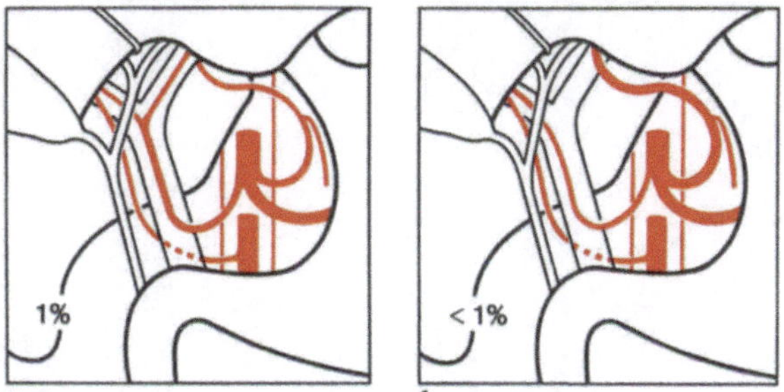

e. f.

g. replaced right hepatic artery from the superior mesenteric and accessory left hepatic artery from the left gastric artery
h. replaced right hepatic artery from the superior mesenteric and replaced left hepatic artery from the left gastric artery

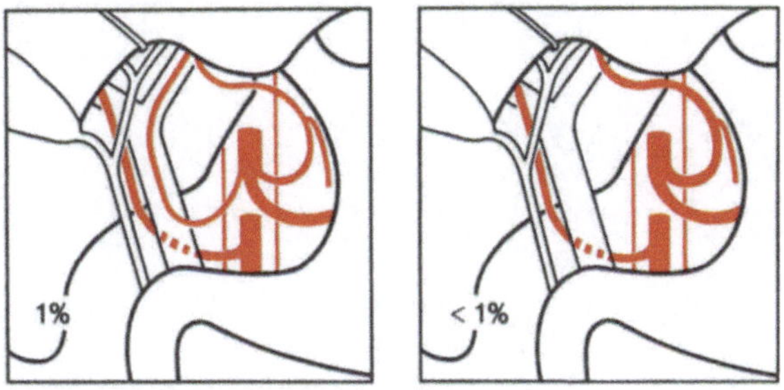

g. h.

15 Cystic artery

The cystic artery normally divides into a superficial branch to supply the ventral side of the gallbladder and a deep branch which supplies the part of the gallbladder adjacent to the liver. This division is usually at the border between the neck and the body of the gallbladder, but in rare cases it is more proximal, occurring as a "high division". The deep branch generally also supplies parts of the liver with branches which are sometimes surprisingly large and probably play an important role in the arterial blood supply of the liver. Data on the frequency of two cystic arteries vary considerably. The reason for this discrepancy might be that the deep branch often leaves the

1. One cystic artery [80%]

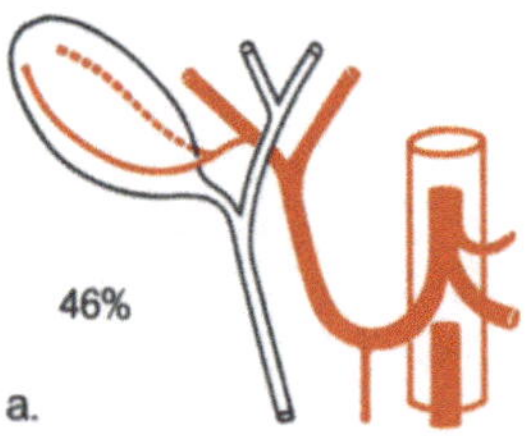

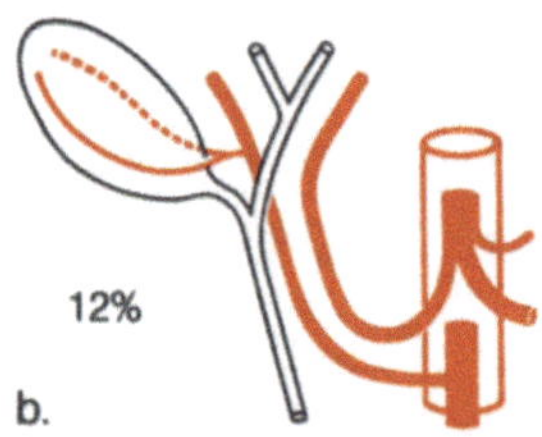

a. cystic artery as a branch of the right hepatic artery
b. cystic artery as a branch of an accessory right hepatic artery which originates from the superior mesenteric artery

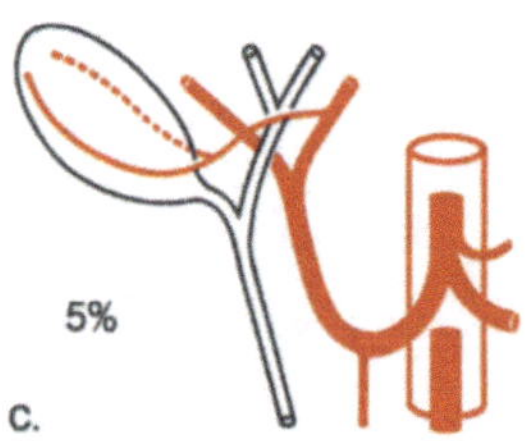

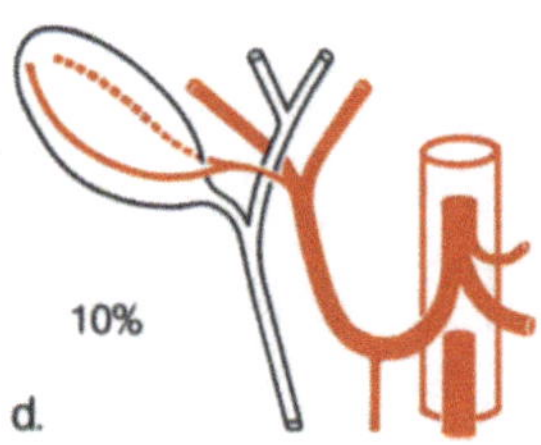

c. cystic artery as a branch of the left hepatic artery
d. cystic artery originates from the bifurcation of the common hepatic artery

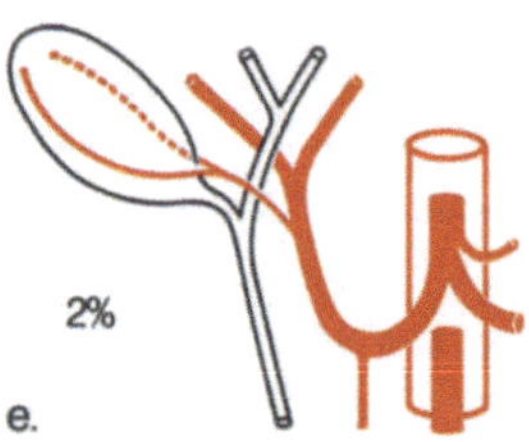

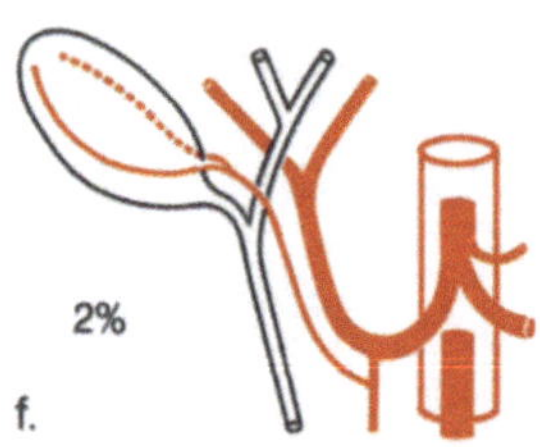

e. cystic artery originates from the proper hepatic artery or very rarely from the right gastric artery
f. cystic artery arises from the gastroduodenal artery or one of its branches

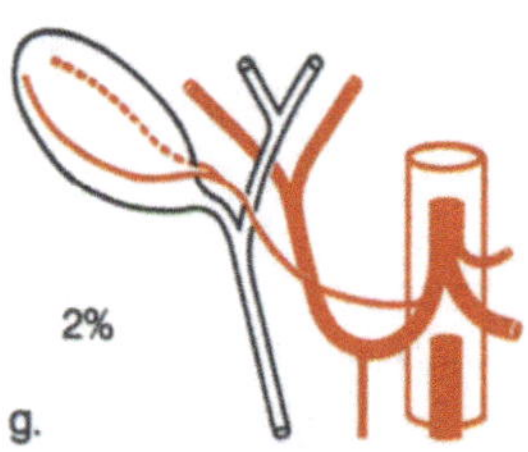

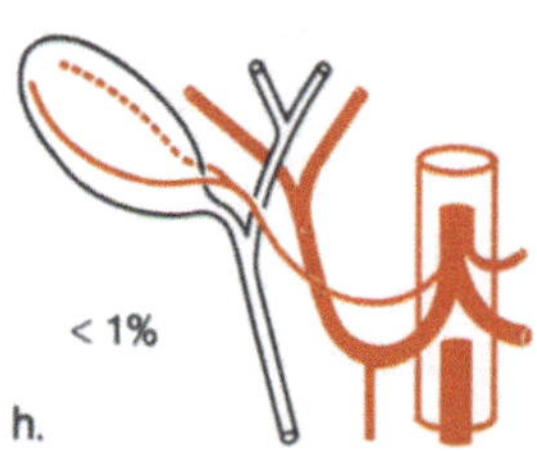

g. cystic artery originates from the common hepatic artery
h. cystic artery arises from the celiac trunk

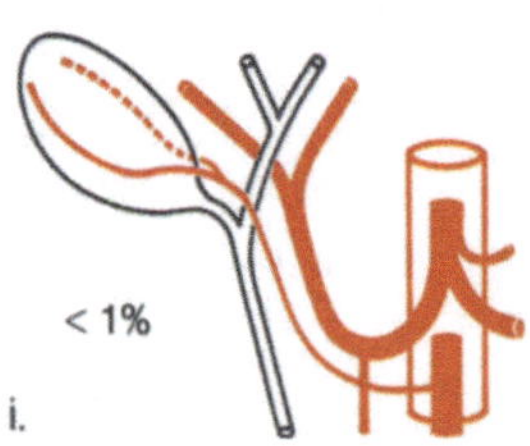

i. cystic artery as a branch of the superior mesenteric artery

right hepatic artery near the liver and can be easily overlooked. If there are two cystic arteries, one artery is nearly always a branch of the right hepatic artery.
About 3/4 of all cystic arteries originate in Calot's triangle, which is formed by the cystic duct, the common hepatic duct and the liver. The remaining 1/4 crosses the biliary ducts, usually at the common hepatic duct or less frequently at the choledochus or right hepatic duct. These arteries normally cross in front of the biliary ducts similar to the surgeon's path of access to this area. If there are two separate cystic arteries, the crossing of the ducts is quite common. If the surgeon detects a cystic artery which crosses the biliary ducts, he should also look for a second cystic artery in Calot's triangle [3, 15*, 40*, 114*, 175, 227, 399a*, 633*].

2. Two cystic arteries 20%

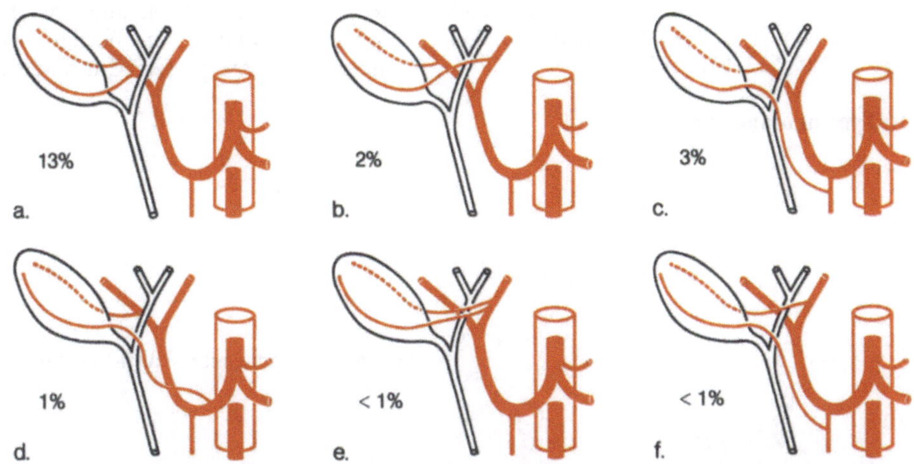

a. both cystic arteries originate from the right hepatic artery, which itself is in ⅓ of all cases a branch of the superior mesenteric artery
b. one cystic artery arises from the right hepatic, one from the left hepatic artery
c. one cystic artery originates from the right hepatic, one from the gastroduodenal artery
d. one cystic artery arises from the right hepatic, one from the common hepatic artery
e. both cystic arteries branch from the left hepatic artery
f. one cystic artery originates from the left hepatic artery, the other from the gastroduodenal artery or one of its branches

3. Three cystic arteries < 1%

4. Topography of the cystic arteries

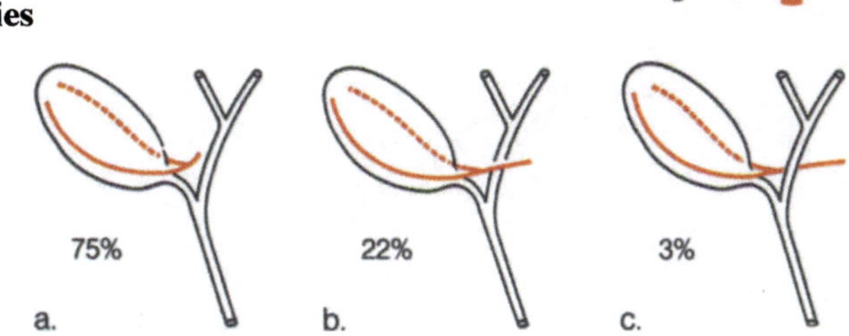

a. cystic artery originates in Calot's triangle
b. cystic artery crosses in front of the hepatic duct
c. cystic artery crosses behind the hepatic duct

16 Splenic artery

Along the course of the splenic artery from the celiac trunk, usually its largest branch, four subdivisions are apparent: suprapancreatic, pancreatic, prepancreatic and prehilar parts. The second segment, the pancreatic part, is in about 90% of all cases suprapancreatic, in 8% behind and in 2% in front of the pancreas. The splenic artery is characterized by a tortuosity which can equal a length of up to 50 cm, although the distance from the aorta to the spleen is only about 10 cm [1, 2, 3, 81, 102, 200, 222, 254, 256, 399 a*, 552, 586, 595, 613].

The origin from the celiac trunk and its variability is shown on pp. 30, 31. In the majority of cases the splenic artery divides into at least two main branches at some distance from the spleen – "distributed type"; the other alternative branches near to the splenic parenchyma – "magistral type". Six to 36 individual arteries enter the spleen without major anastomoses within the spleen. These arteries supply the segment-like parts of the spleen, and are of great clinical importance in partial splenectomies, which are preferred nowadays to a

1. Type of branching

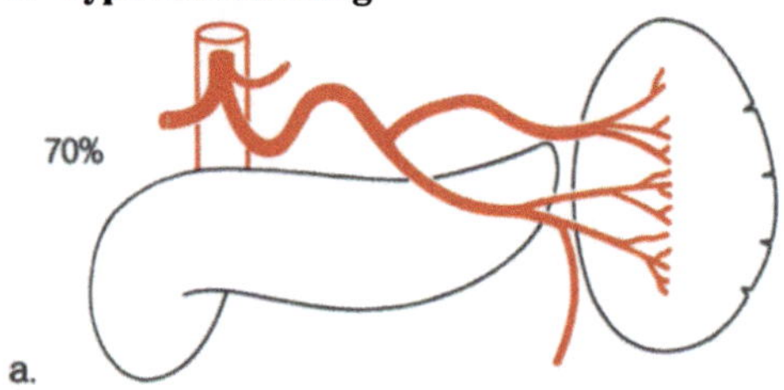

a. remote from the hilus – distributed type

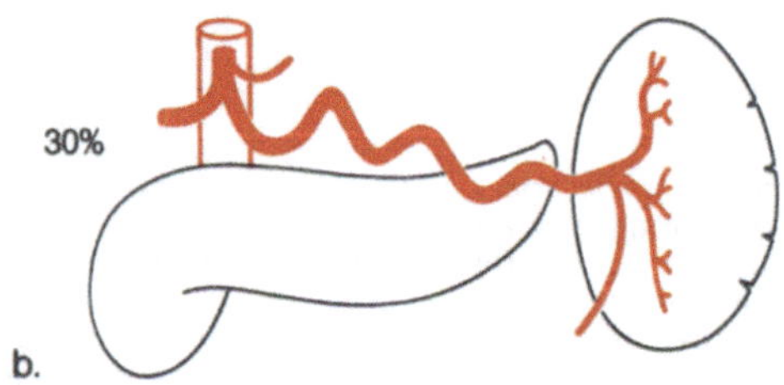

b. near the hilus – magistral type

2. Polar arteries

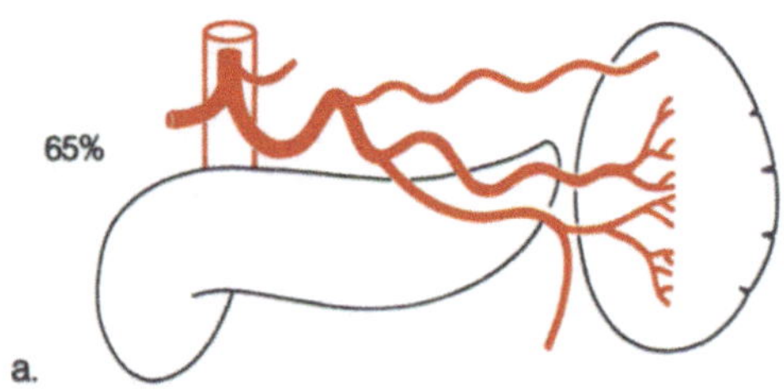

a. upper polar artery

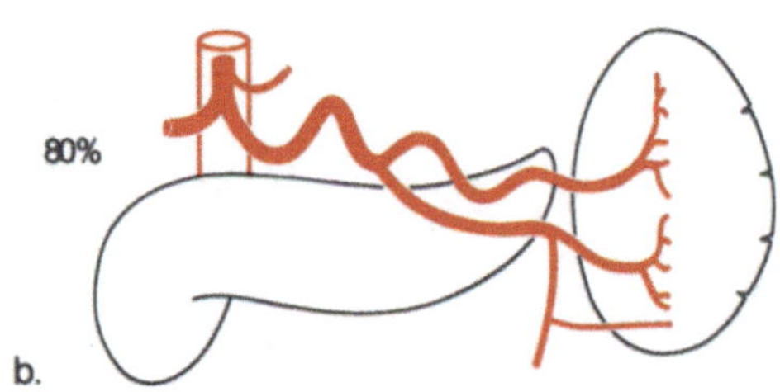

b. lower polar artery

total splenectomy [222, 254, 291, 399a, 644].
Important branches of the splenic artery are the short gastric arteries and the left gastroepiploic artery (see gastric arteries, p. 40). Arteries to the cranial or caudal pole of the spleen originate from the main splenic artery or the superior or inferior terminal branches. In rare instances, the upper polar arteries arise from the aorta, the left gastric artery, a pancreatic or even a mesenteric artery [3, 307, 399a].

3. Position of the splenic artery

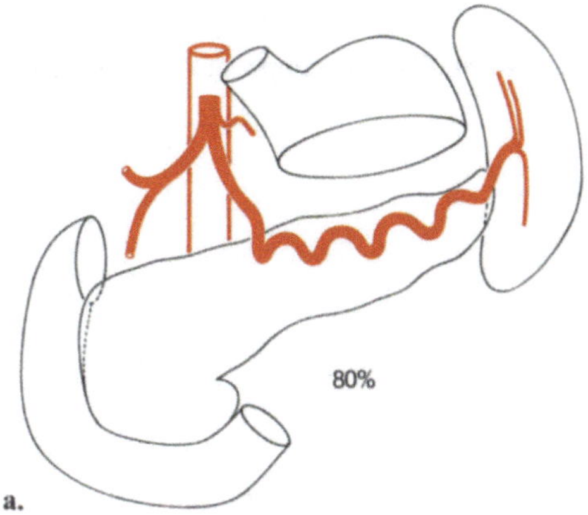

a. along the upper border of the pancreas (suprapancreatic)

a.

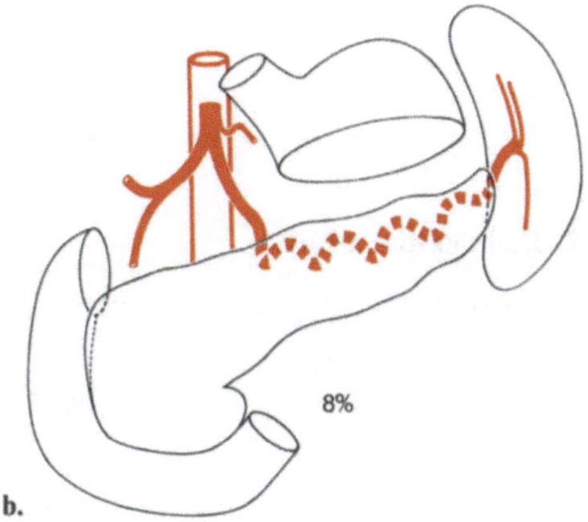

b. behind the pancreas (retropancreatic)

b.

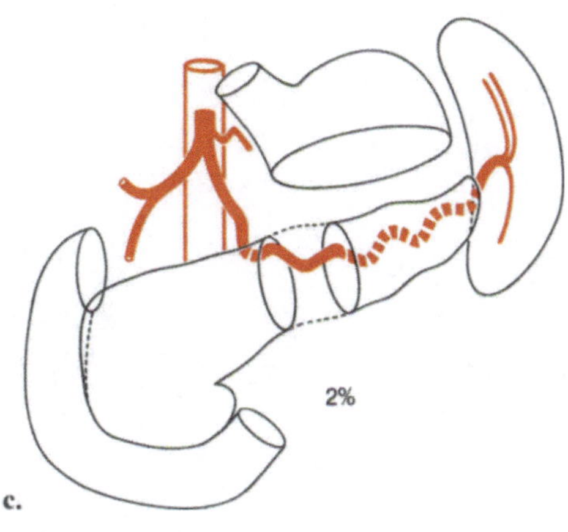

c. within the pancreas (intrapancreatic)

c.

17 Gastric arteries

Textbooks usually list six arteries for the stomach and outline them in figures. The left and right gastric arteries anastomose along the lesser curvature. The left and right gastroepiploic arteries are found along the greater curvature. The fundus of the stomach is supplied by the short gastric arteries, which vary in number, and the pyloric region, by the gastroduodenal artery. The arterial anastomoses at the cardio-oesophageal (see oesophageal arteries, p. 17) and at the gastroduodenal end of the stomach are extremely variable. In most cases the left gastric artery is more promi-

1. "Normal" blood supply of the stomach

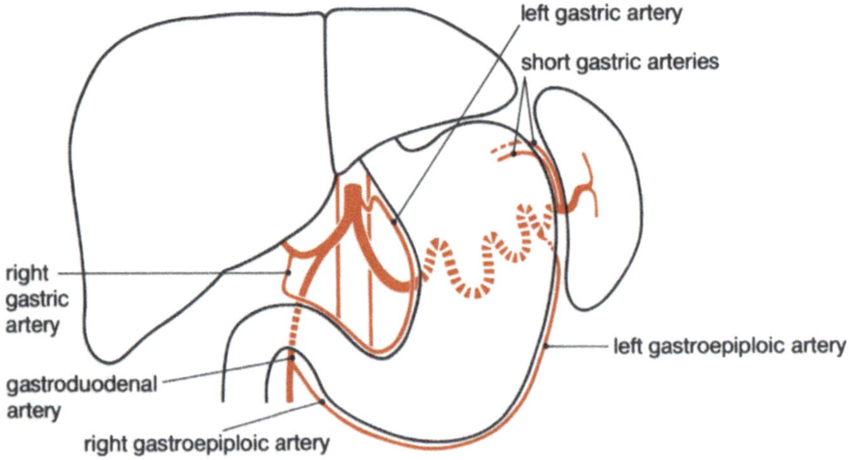

2. Right gastric artery

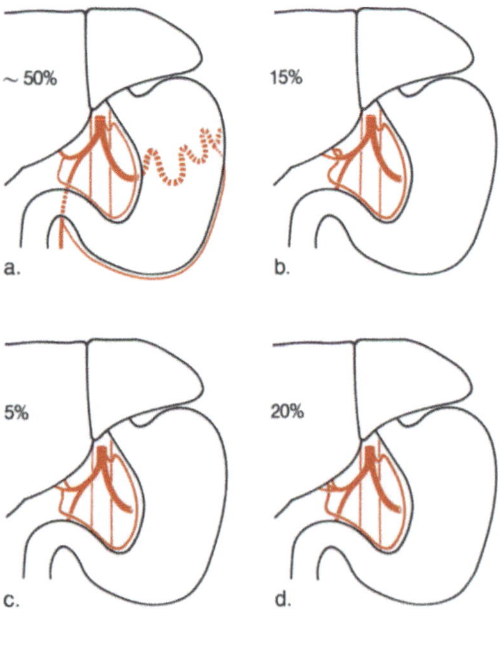

a. "normal" situation, origin from the common hepatic artery
b. origin from the left hepatic artery

c. origin from the right hepatic artery
d. origin from a middle hepatic artery or at the bifurcation of the proper hepatic artery

nent than the right. At the greater curvature, however, the right gastroepiploic artery is nearly always larger than the left, and the right gastroepiploic artery usually ends beyond the midline of the lower border [3, 18, 156, 167, 263, 339, 347*, 348*, 399a*, 403*, 406, 453, 495*, 515, 525, 528, 633]. The gastroduodenal artery usually has a pyloric branch and is important as the origin of the right gastroepiploic artery. Origins from the common hepatic trunk or left, right or middle hepatic arteries, or directly from the superior mesenteric artery have been described [167, 175, 270, 367, 394, 406, 477, 483, 495, 528, 668].

3. Left gastric artery

The left gastric artery arises in about 90% of all cases from the celiac trunk and is usually the first branch. It is of great importance that in about one of four patients the left gastric artery participates in the blood supply of the liver, and in half of these cases it is an accessory hepatic but in the other half (11.5%) a replaced hepatic artery. The left gastric artery usually divides into two main branches, anterior and posterior [3, 156, 167, 348, 399a*, 428, 456, 596*]. The right gastric artery anastomoses in most cases with the posterior branch of the left gastric artery [3, 141, 156, 348, 399a].

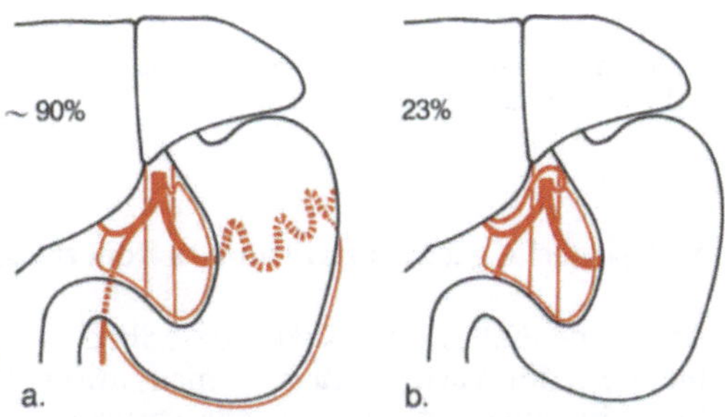

a. "normal" origin from the celiac trunk
b. an accessory hepatic artery arises from the left gastric artery

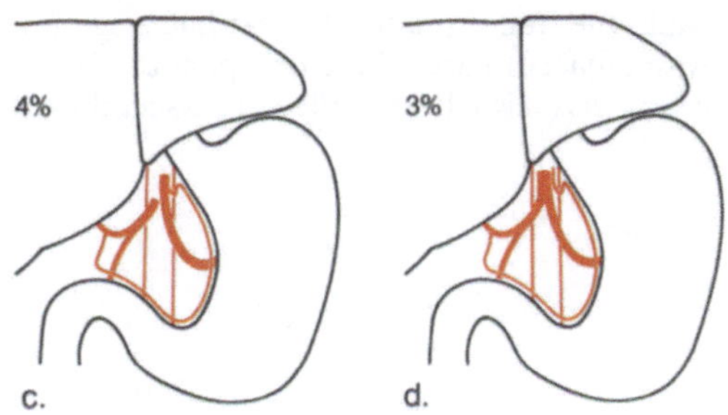

c. origin from a lienogastric trunk
d. origin direct from the aorta

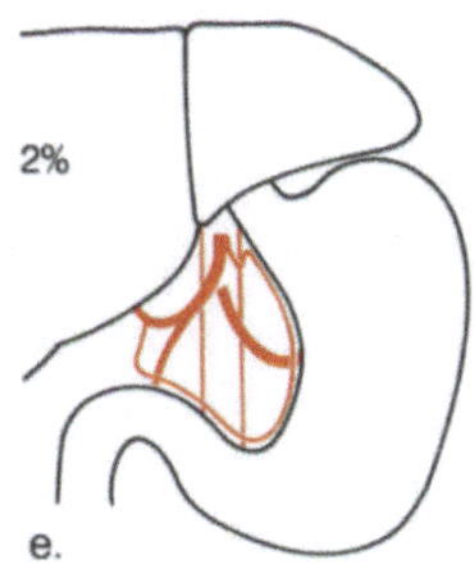

e. origin from a hepatogastric trunk

4. Gastroepiploic arteries

The left gastroepiploic artery is highly variable in origin, size, length and number of branches. It generally originates from the splenic trunk or from an inferior splenic terminal, but sometimes from the superior terminal or even the interior of the spleen [3, 156, 348, 399 a]. Of clinical importance is its origin, either directly or with other arteries, from the superior mesenteric artery. The gastroepiploic arch can be near to the stomach wall or as far as 2 cm from the greater curvature [3, 156, 241, 348, 399 a].

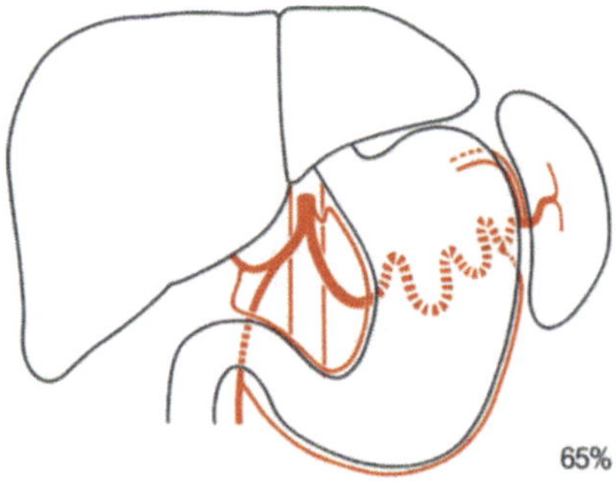

a. "normal" pattern with an anastomosis of the right and left gastroepiploic arteries

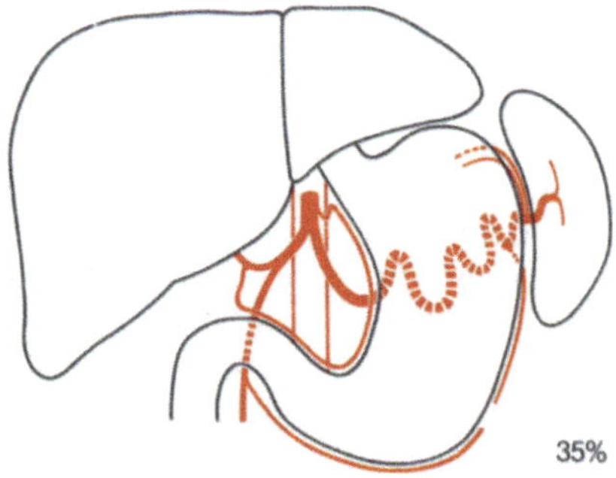

b. no real anastomosis between the right and left gastroepiploic artery

5. A posterior gastric artery derives from the splenic artery 60%

An artery that is not usually described is the posterior gastric artery, which arises from the splenic artery, ascends vertically or obliquely, and supplies the posterior wall, the fundus and the cardiac region with adjacent parts of the oesophagus. This artery has also been called an ascending posterior oesophagogastric branch or accessory left gastric artery. The term posterior gastric artery should now be used since this name is included in the Nomina Anatomica (a. gastrica posterior) [3, 122*, 399 a, 598*].

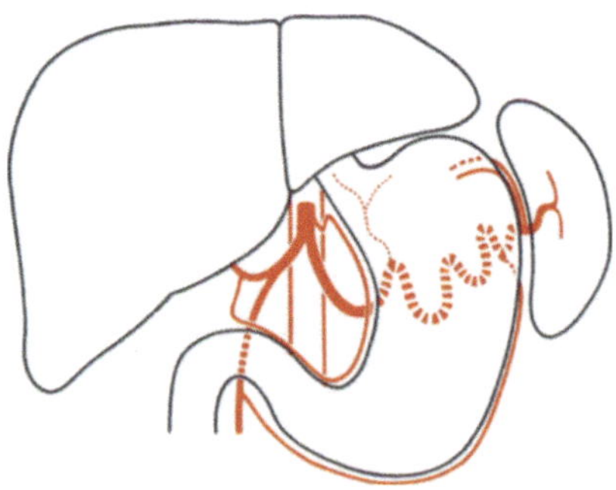

18 Pancreatic arteries

The arterial supply of the pancreas is highly variable, but follows a general pattern [60, 64, 65, 97, 176, 202a*, 203, 255, 270, 394, 394, 399a*, 403, 406, 454*, 477, 483*, 495, 504, 510*, 523, 552, 628, 645, 668*]. The head of the pancreas is supplied by anastomosing branches from the celiac trunk and the superior mesenteric artery. These arteries form arcades in front and behind the head of the pancreas on the vertical axis. The body of the pancreas is supplied by arteries running in the longitudinal direction of the pancreas. The cauda of the pancreas is supplied by numerous branches from the splenic artery [176, 552], which are not shown in the figures.

In selective pancreatic angiography the number of arteries and branches identified largely depends on the position of the catheter and the direction of the blood flow in the different anastomosing arteries around or within the pancreas. Fewer pancreatic arteries are detected than in anatomical studies. The large pancreatic artery (a. pancreatica magna) has not been depicted in detail in the figures.

The anterior and posterior arcades of the head of the pancreas vary not only in number but also in origin. The anterior superior pancreaticoduodenal artery is very constant representing the terminal branch of the gastroduodenal artery.

The superior origin of the posterior arcade varies more often (Fig. 5). Both inferior origins of the anterior and posterior arcades leave the superior mesenteric artery as a short common trunk (~65%). The other alternative origins for the anterior arcade are shown in Fig. 3.

The transverse pancreatic artery to the body of the pancreas is connected with the dorsal pancreatic and the large pancreatic arteries. The transverse pancreatic artery is often described as the inferior pancreatic and the (less common) superior pancreatic arteries. This might partly explain the considerable discrepancy in the literature on the frequency of this artery. In addition to Fig. 7–c other origins, such as the common hepatic, gastroduodenal, and the dorsal arcade, have been described.

1. Pancreatic arteries as shown in textbooks

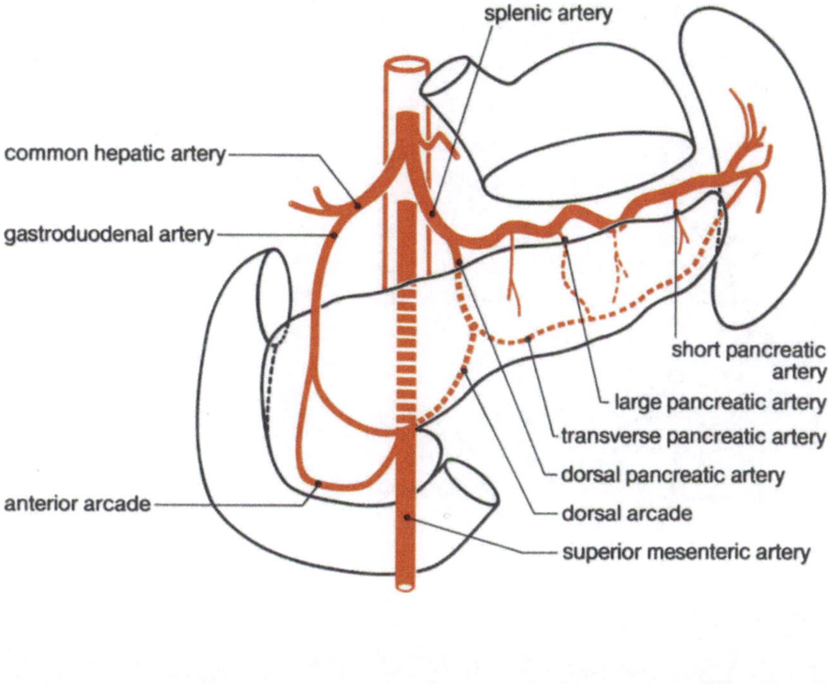

2. Arterial arcades in front of the head of the pancreas

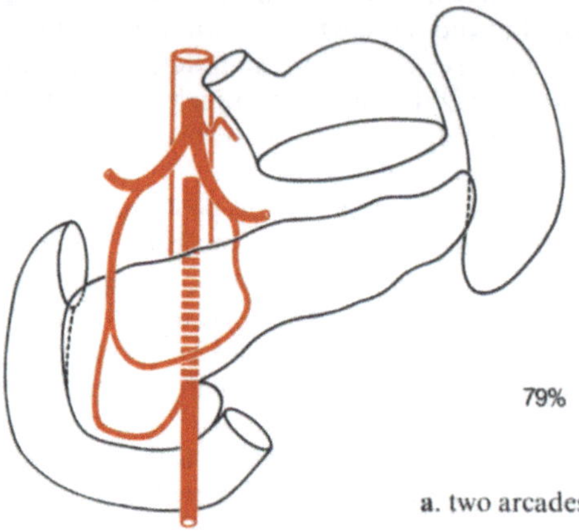

a. two arcades

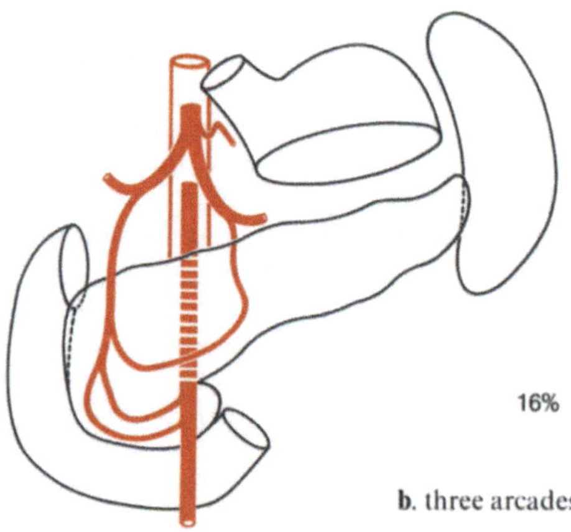

b. three arcades

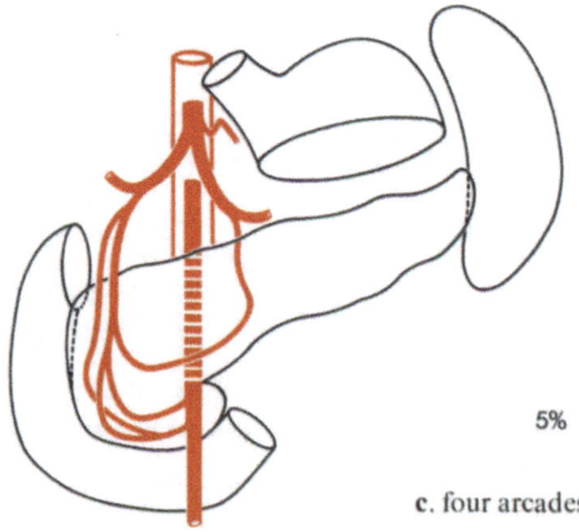

c. four arcades

3. Inferior origin of the anterior arcades

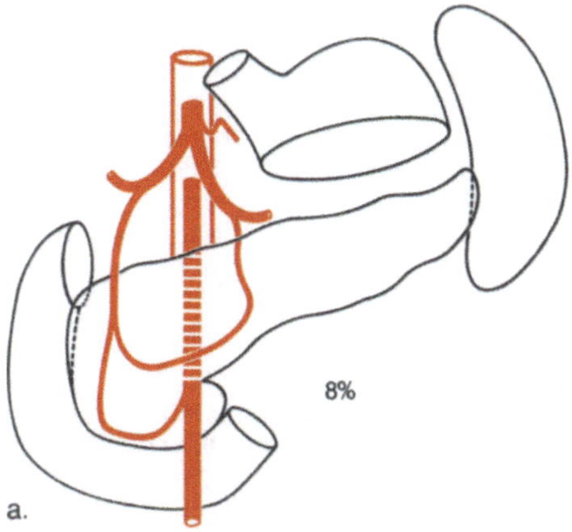

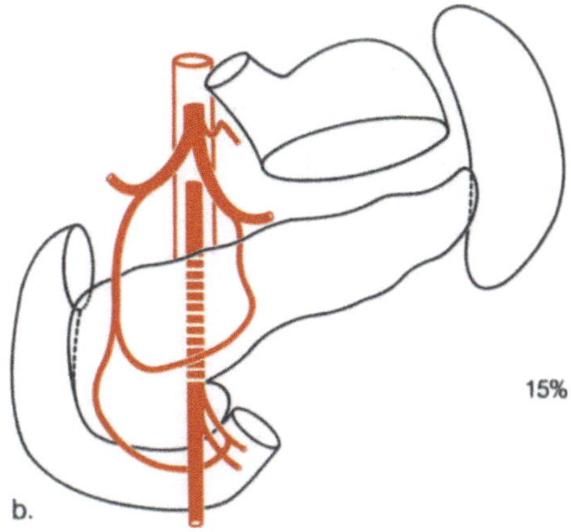

a. from the right side of the stem of the superior mesenteric artery
b. from the first jejunal artery

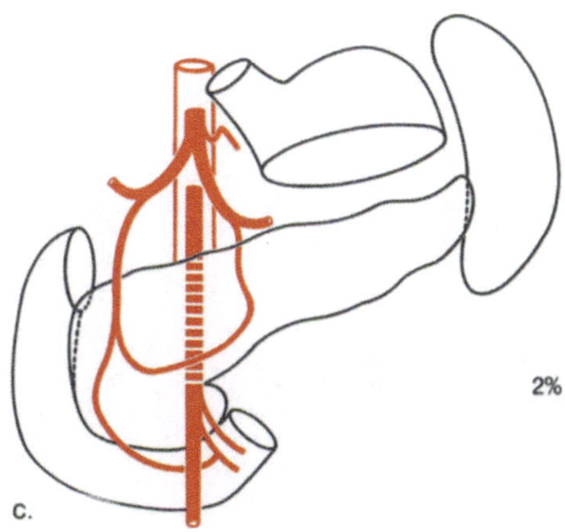

c. from the second jejunal artery

4. Dorsal arterial arcades

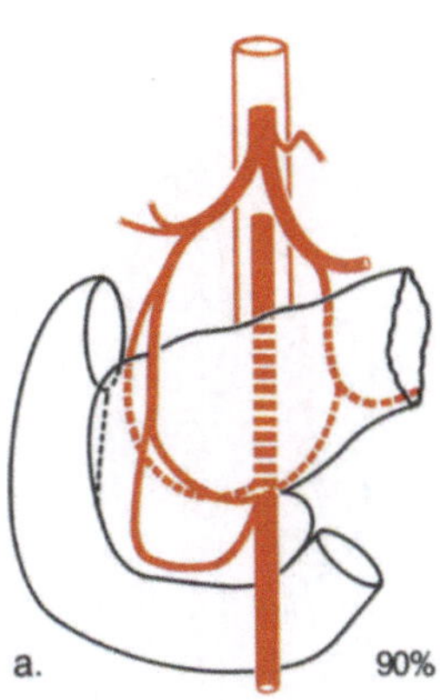

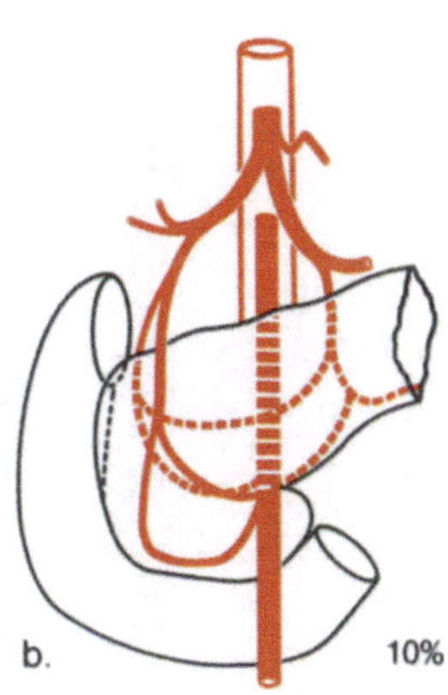

a. one arcade
b. two arcades

5. Superior origin of the posterior arcade

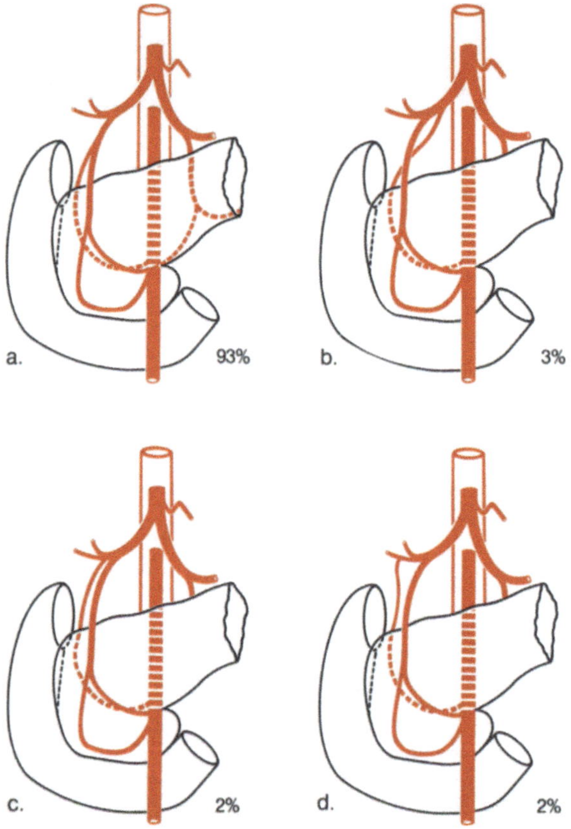

a. from the gastroduodenal artery
b. from the common hepatic artery

c. from the proper hepatic artery
d. from the right hepatic artery

6. Origin of the dorsal pancreatic artery

a. splenic artery
b. celiac trunk
c. common hepatic artery

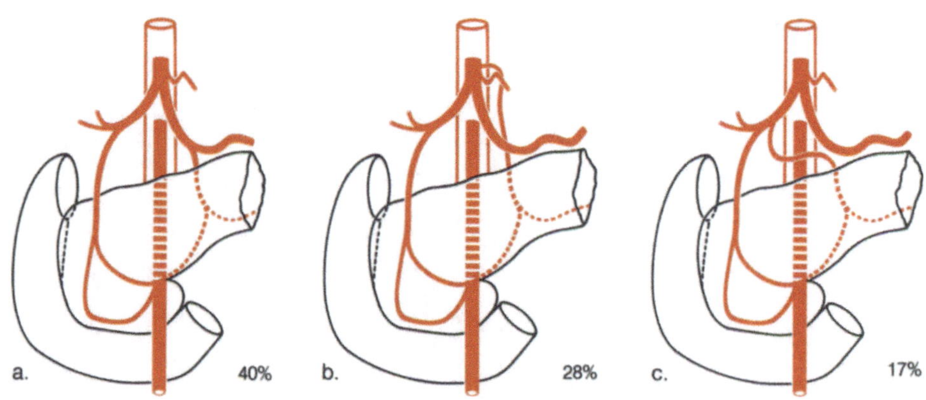

d. superior mesenteric artery
e. from an aberrant hepatic artery from the superior mesenteric artery

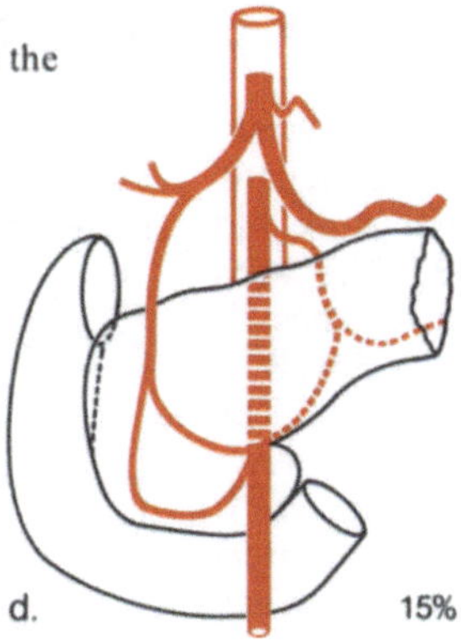

d. 15%

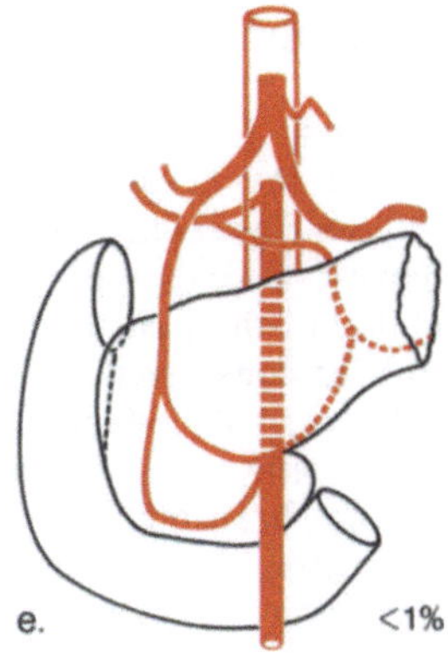

e. <1%

7. Origin of the transverse pancreatic artery

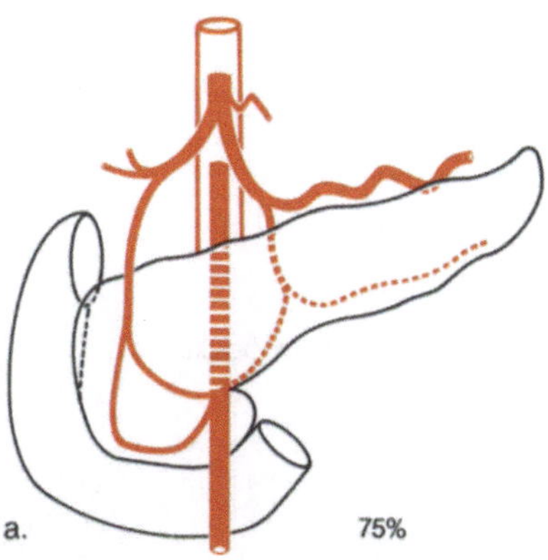

a. 75%

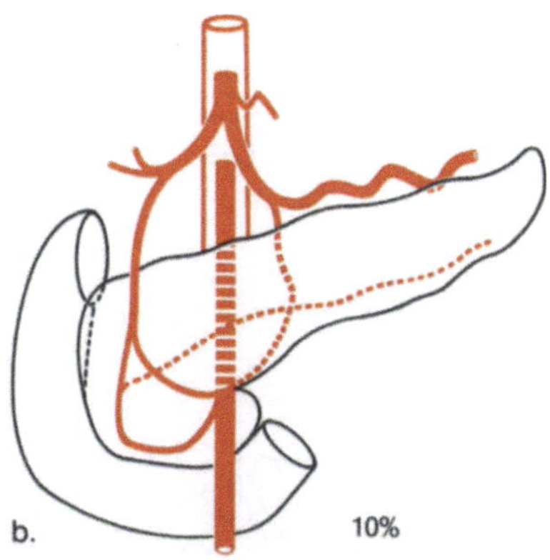

b. 10%

a. dorsal pancreatic artery
b. anterior arcade

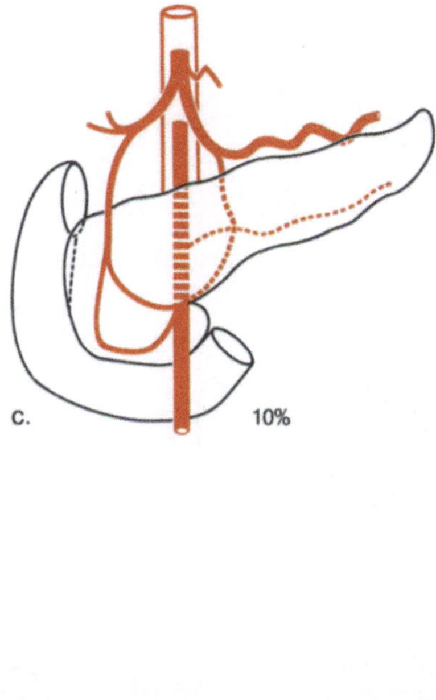

c. 10%

c. superior mesenteric artery

19 Superior mesenteric artery and celiac trunk

The basis for these figures are the anomalies of the celiac trunk and the hepatic arteries (for references see celiac trunk and hepatic arteries, pp. 30, 32–33). The variability of the gastroduodenal artery has also been considered [167, 175, 270, 367, 394, 406, 477, 483, 495, 525, 528, 668]. The combination of different accessory arteries has not been included. From the figures it is obvious that there is no clearcut watershed between organs supplied by the celiac trunk and those supplied by the artery of the embryological mid-gut, the superior mesenteric artery.

- primarily supplied by the superior mesenteric artery (solid shading)
- partly supplied by the superior mesenteric artery, partly by branches of the celiac trunk (dotted shading)
- primarily supplied by branches of the celiac trunk (unshaded)

1. Blood supply of the superior mesenteric artery to abdominal organs

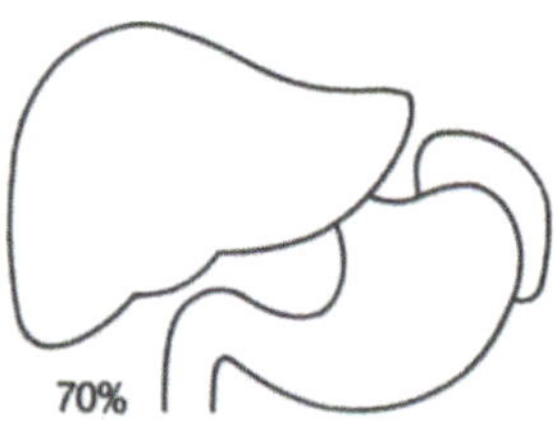

a.

a. "normal" type as shown in textbooks

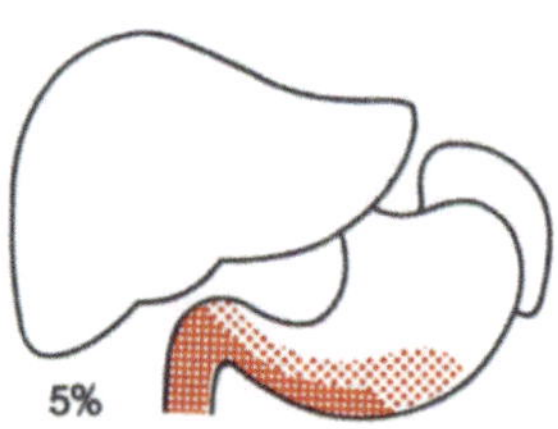

b.

b. gastroduodenal artery from the superior mesenteric artery

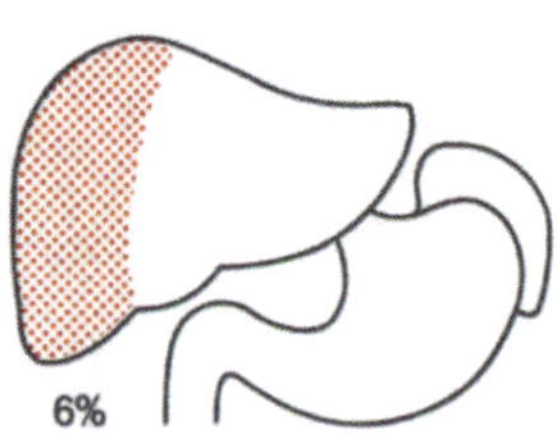

c.

c. accessory right hepatic artery from the superior mesenteric artery

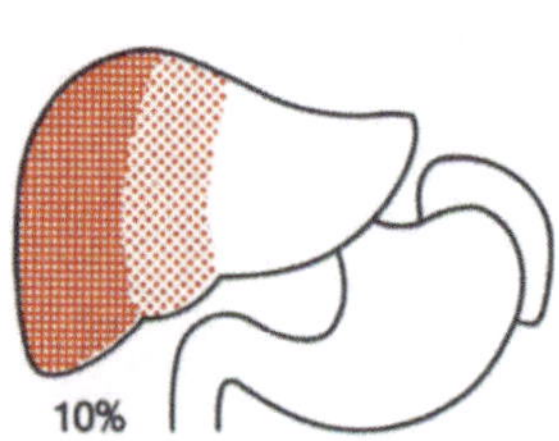

d.

d. right hepatic artery from the superior mesenteric artery

e. right hepatic and gastroduodenal arteries from the superior mesenteric artery

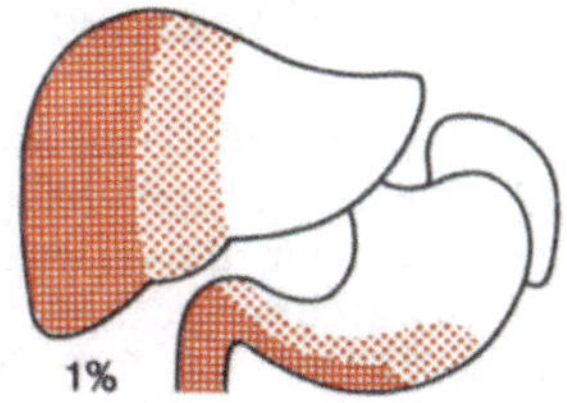

e.

f. proper hepatic artery from the superior mesenteric artery

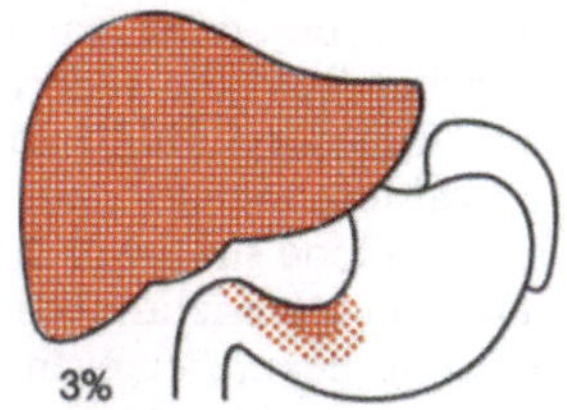

f.

g. common hepatic artery (hepatomesenteric trunk) from the superior mesenteric artery

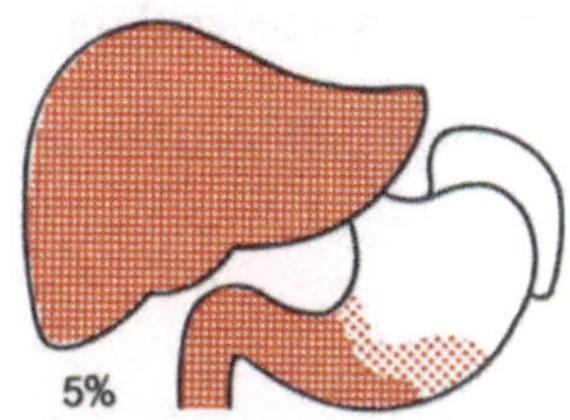

g.

h. common hepatic and splenic arteries form a common trunk with the superior mesenteric artery (hepatosplenomesenteric trunk)

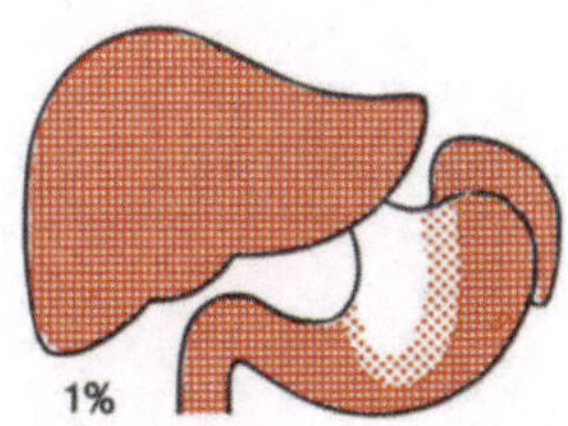

h.

i. celiac trunk and superior mesenteric artery have a common origin (gastrohepatosplenomesenteric trunk)

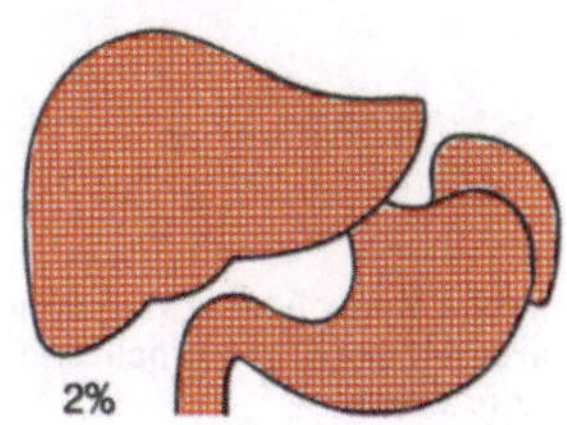

i.

k. splenic artery originates with the superior mesenteric artery (splenomesenteric trunk)

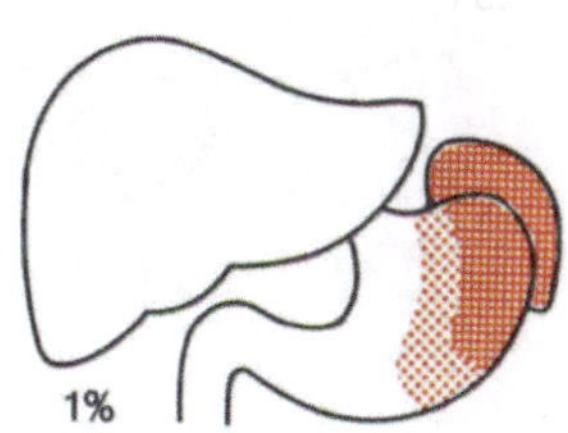

k.

20 Superior mesenteric artery and colic arteries

The gastroduodenal arteries (see pp. 32, 38, 39), the jejunal and ileal arteries are not depicted, but the arteries supplying the colon, caecum and appendix are outlined in detail [15*, 18, 65*, 96, 159, 184, 218, 263, 270, 399 a*, 403, 435 a, 453*, 495*, 525, 528, 582*, 587, 655, 676]. The ileocolic artery is the most variable of the colic branches from the superior mesenteric artery. Accessory middle colic arteries are of clinical importance, since in such cases the arterial blood supply of the colon by the superior mesenteric artery is extended to the left colic flexure or even as far as the descending colon. Accessory middle colic arteries have been described as branches of the gastroduodenal or left gastroepiploic arteries. If accessory branches from the inferior mesenteric artery are present, arteries supply the transverse colon up to the right colic flexure. Normally the anastomosis along the colon formed by different arches (marginal artery) is large enough to ensure a sufficient collateral circulation (anastomosis of Riolan). This anastomosis between the superior and inferior mesenteric arteries is absent in about 2% of all cases and in a further 3% very small. The arch between the right colic and the ileocolic arteries seems to be absent in about 6% of all cases [15*, 18, 32, 36, 65, 88, 98, 242, 366, 399 a, 435 a, 570, 571, 572, 573, 582*, 593, 631, 632, 635].

1. Three colic arteries branch from the superior mesenteric artery 67%

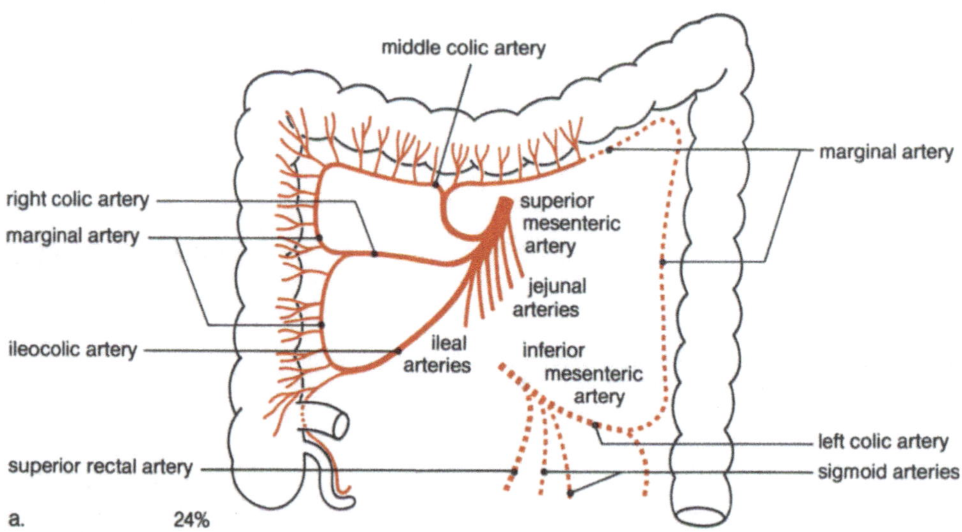

a. the ileocolic, right and middle colic arteries are independent branches of the superior mesenteric artery
b. the ileocolic and right colic arteries form a trunk
c. the right and middle colic arteries form a common trunk
d. the ileocolic, right and middle colic arteries form a common trunk

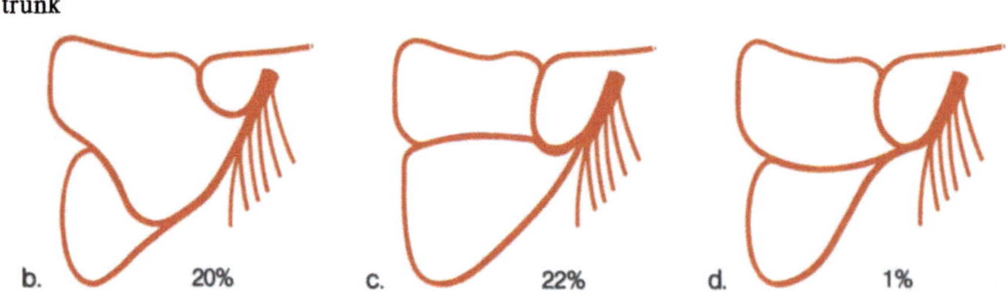

2. Two colic arteries branch from the superior mesenteric artery [15%]

a. the right colic artery is absent; the ileocolic and middle colic arteries in such cases normally branch separately from the superior mesenteric artery
b. the middle colic artery is absent; the ileocolic and right colic arteries branch separately in ¾ of all cases; otherwise they form a common trunk

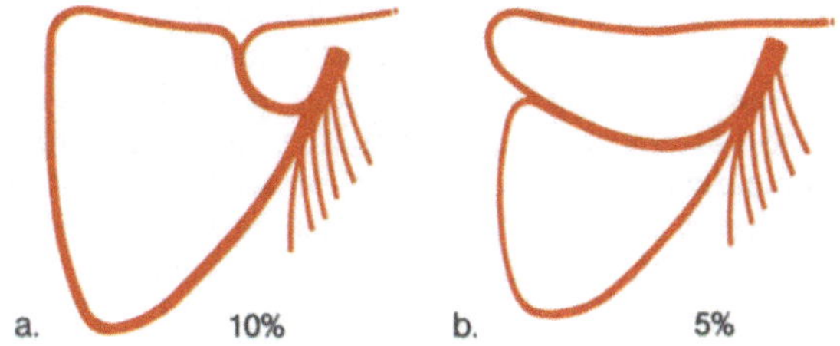

3. Accessory colic arteries [18%]

a. two right colic arteries normally from common trunks with the middle colic and ileocolic artery
b. three right colic arteries

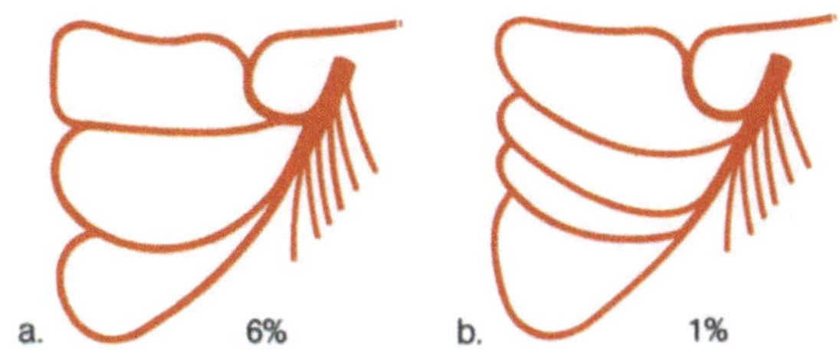

c. two middle colic arteries; all four colic arteries originate separately
d. two middle colic arteries with a common trunk of the ileocolic and right colic arteries

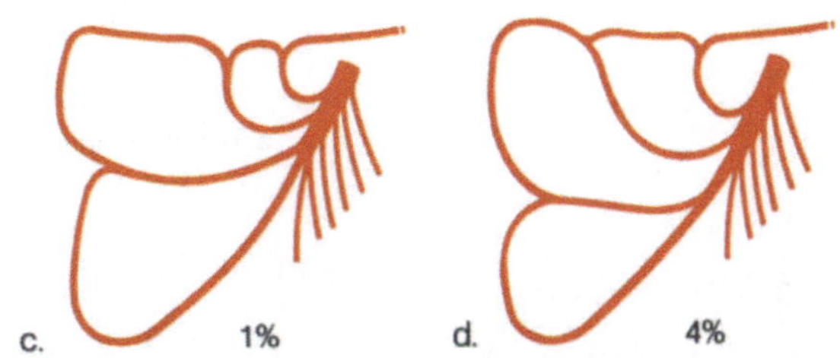

e. two middle colic arteries with a common trunk of one middle colic artery and the right colic artery
f. three middle colic arteries

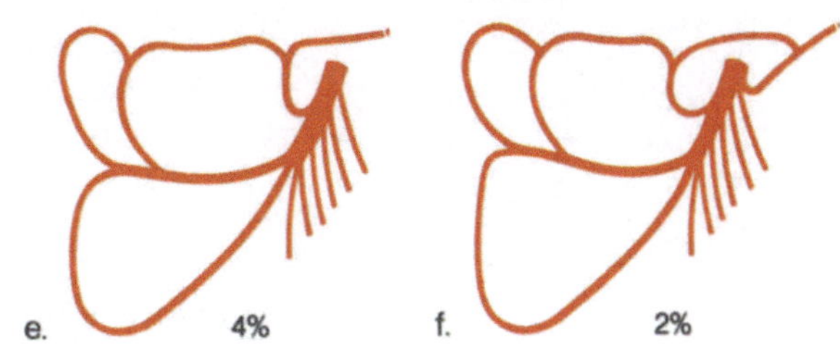

4. Left colic artery is a branch of the superior mesenteric artery [<1%]

a. left accessory colic artery from the superior mesenteric or middle colic arteries
b. main stem of the left colic artery originates from the superior mesenteric artery (sometimes from a common trunk with the middle colic artery)

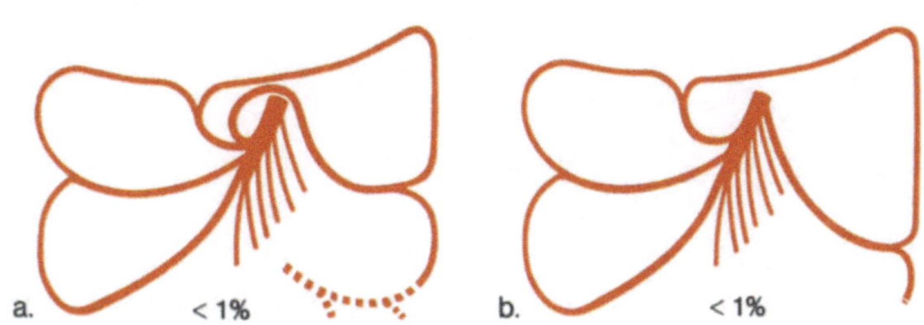

21 Appendicular artery

The figures showing the appendicular artery are very schematic because variations of the ascending, anterior, and posterior caecal arteries have not been included. The appendicular artery can run ventrally to the ileum if this artery branches very distally from the caecal artery. Normally even an appendicular artery as a branch of the anterior caecal artery runs dorsally to the ileum. Duplications of the appendicular artery have been described if one artery is dorsal and the other ventral to the ileum [15*, 36, 98*, 384, 435 a, 631*, 635*, 670].

1. Appendicular artery runs dorsally to the ileum >99%

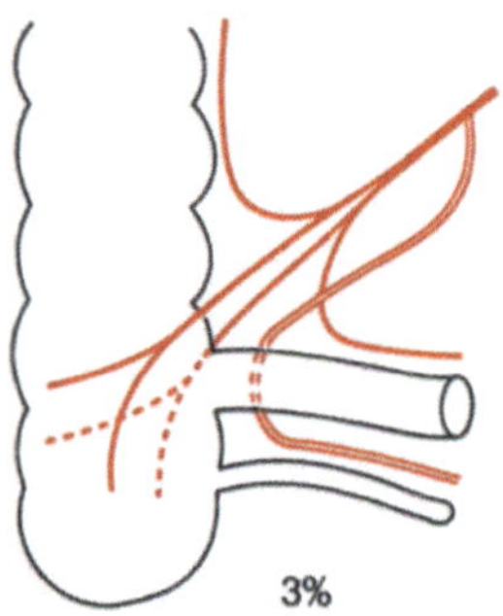

a.

a. origin from the ileocolic artery proximal to its division

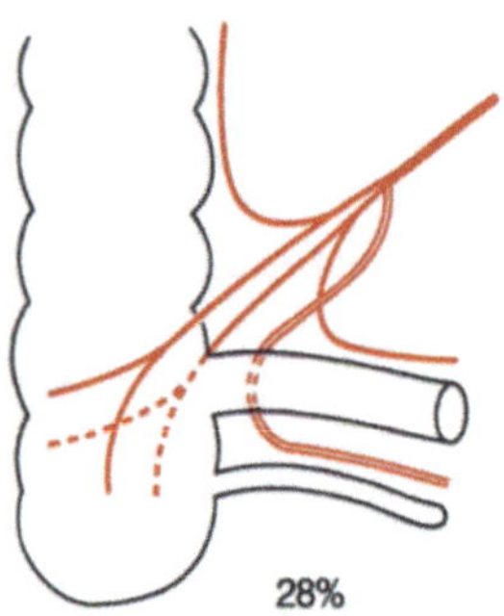

b.

b. origin from the division of the ileocolic artery

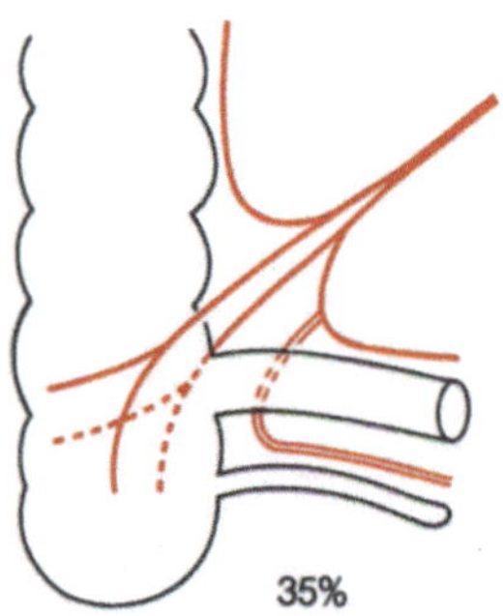

c.

c. origin from the last branch to the ileum

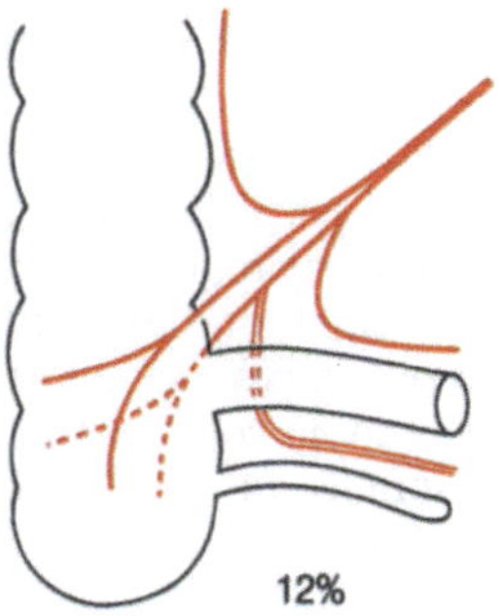

d. origin from a posterior caecal artery

d.

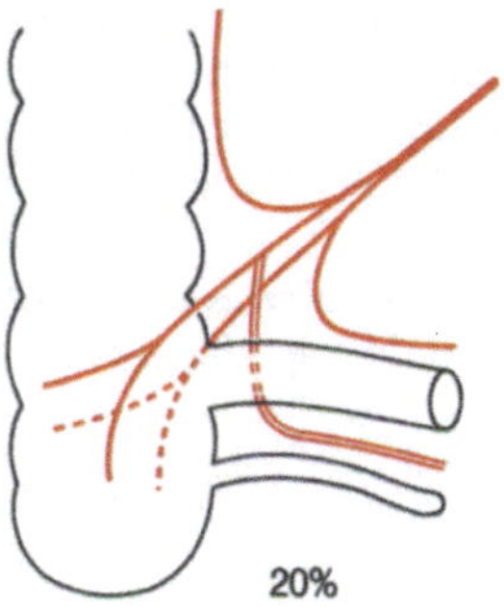

e. origin from the anterior caecal artery

e.

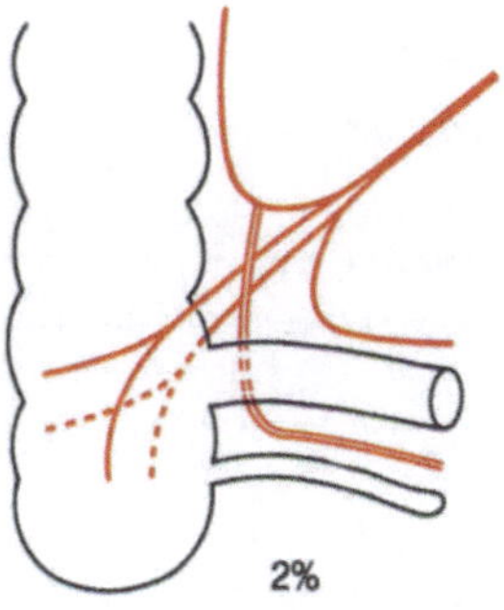

f. origin from the ascending caecal branch or a common trunk of the ileocolic and right colic artery

f.

2. Appendicular artery runs ventrally to the ileum <1%

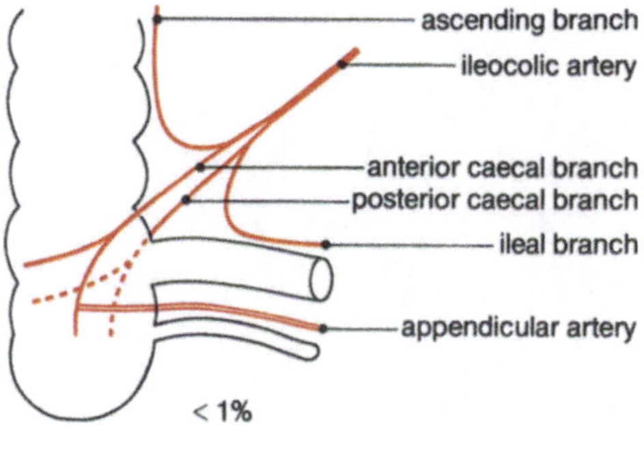

22 Inferior mesenteric artery

The inferior mesenteric artery is an artery with a generally constant origin. It nearly always originates from the aorta, usually at the height of the third lumbar vertebra, supplying parts of the transverse colon, the descending colon, sigmoid and parts of the rectum [3, 14, 15, 26, 42, 51, 142, 159, 224, 225, 242, 287, 318, 366, 384, 395, 402a*, 518, 521, 583, 634]. The number of branches to the sigmoid varies considerably from 1 to 6. For the sake of clarity only two sigmoid arteries are outlined in these figures. The most important end branch of the inferior mesenteric artery is the superior rectal artery, which in 80% of all cases divides into two main branches. These anastomose with the four other rectal arteries, which are branches of the internal iliac artery [26, 659]. The superior rectal artery is the main arterial blood supply to the rectum. Therefore, a ligature distal to the last anastomosis with a sigmoid artery (the point of Sudeck) can be dangerous. These rectal anastomoses can be of clinical importance when the external iliac artery is occluded and the collateral blood supply of the leg uses the inferior mesenteric, superior rectal, internal iliac or femoral arteries. In such cases a mesenteric steal syndrome can be observed [123, 384, 521].

1. The inferior mesenteric artery branches into the left colic artery, sigmoid arteries and the superior rectal artery 89%

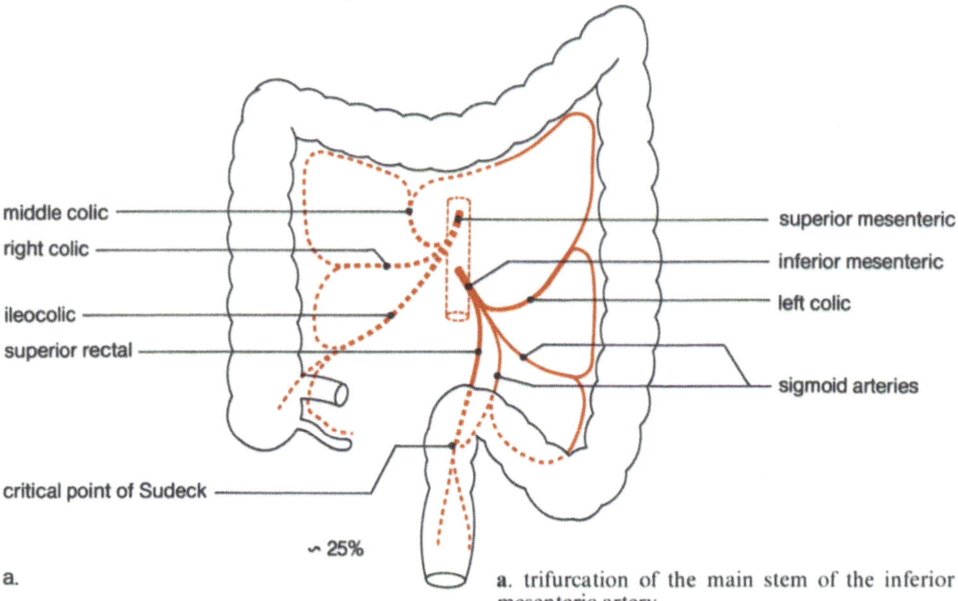

a. a. trifurcation of the main stem of the inferior mesenteric artery

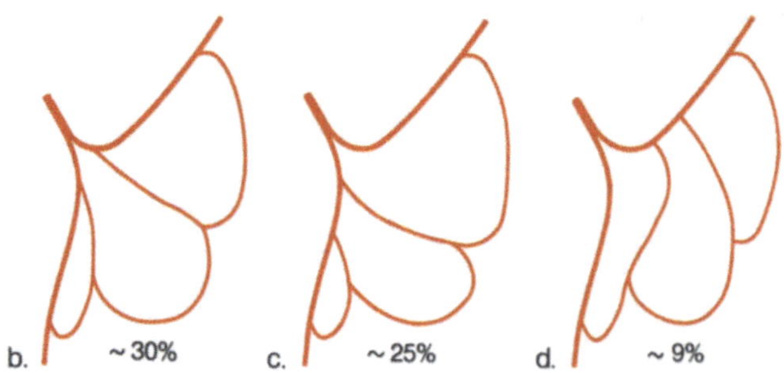

b. bifurcation of the main stem; both stems give off branches to the sigmoid
c. bifurcation of the inferior mesenteric artery; the sigmoid arteries originate only from the superior rectal artery
d. bifurcation of the main stem; the sigmoid arteries originate mainly from the left colic artery

It is difficult to define an exact border between the areas supplied by the superior and inferior mesenteric arteries because of the arterial arches, the marginal arteries along the colon. In anatomical studies (arterial dissections) the border is normally delimited more orally than in angiographic studies. Angiography shows the functional situation better but gives evidence only for the area supplied by the dye-injected artery and does not exclude an additional supply by another artery. Therefore, the figure is a compromise between anatomical and angiographic data [15, 123, 215, 366, 381, 572, 573, 582, 593, 632*, 634].

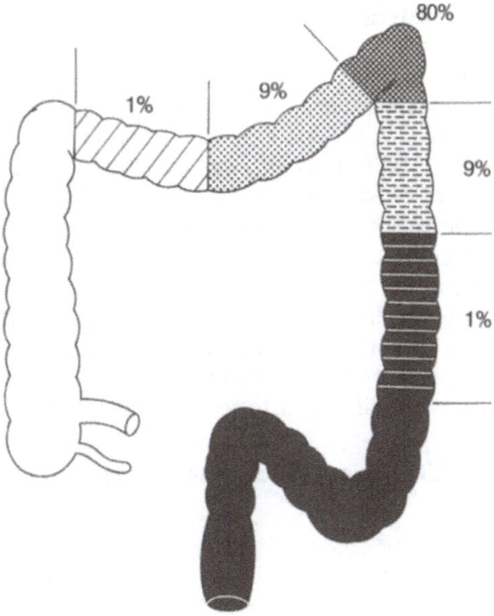

2. Middle colic or accessory middle colic artery originates from the inferior mesenteric artery 10%

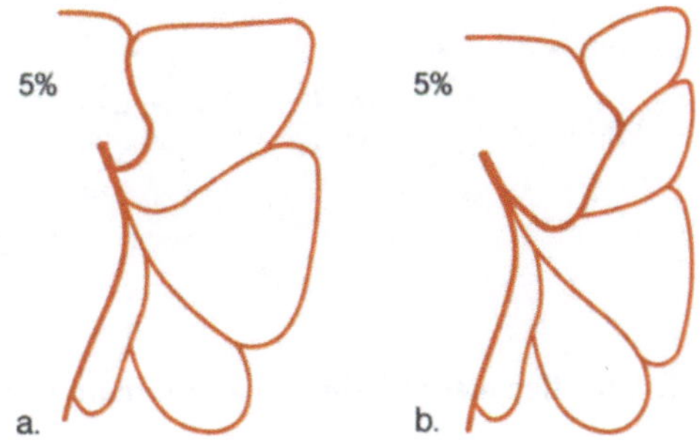

a. (accessory) middle colic artery from the main stem of the inferior mesenteric artery
b. (accessory) middle colic artery from the left colic artery

3. Rare situations 1%

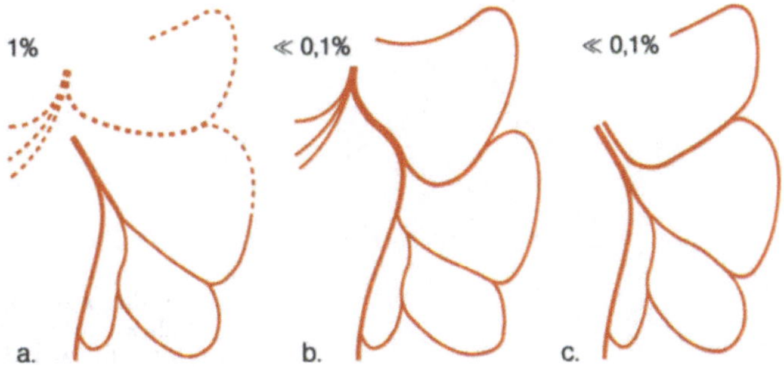

a. left colic artery is absent or originates from the superior mesenteric artery
b. inferior mesenteric artery arises from the superior mesenteric artery
c. two inferior mesenteric arteries with the left colic artery as a direct branch of the aorta

23 Internal iliac artery

The parietal and visceral arteries branch from the internal iliac artery in numerous ways [3*, 15, 24, 173, 356, 358, 431, 441*, 461, 519*]. It is difficult to classify these types according to embryological development or by practical features. According to embryological development, the umbilical artery must be the main branch and the other arteries only side branches [441*]. After birth the umbilical artery is no longer important, and in about 0.7% of all cases one umbilical artery is absent. These anomalies are not shown in the figures. Some recent studies have described and partly summarized such cases along with the relevant literature [185*, 648*]. The sequence of branches or trunk formations is extremely variable [33, 441*]. This situation is also complicated by the various names of these arteries in the different anatomical nomenclatures. In some clinical studies the internal iliac artery is still called the hypogastric artery.

1. All arteries branch from one main stem of the internal iliac artery 10%

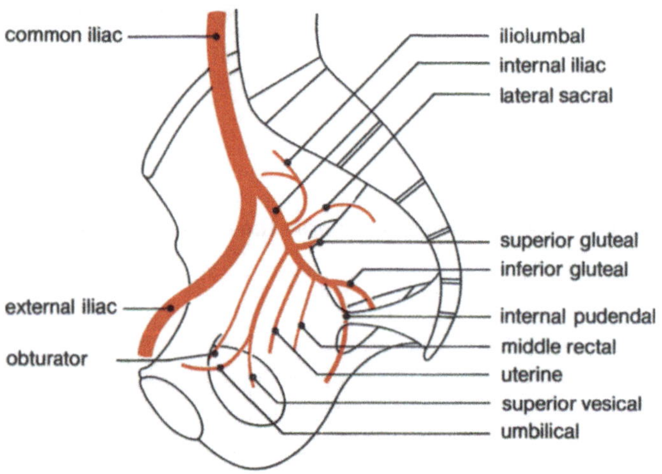

2. The iliac artery divides into two main stems which give off the other branches 60%

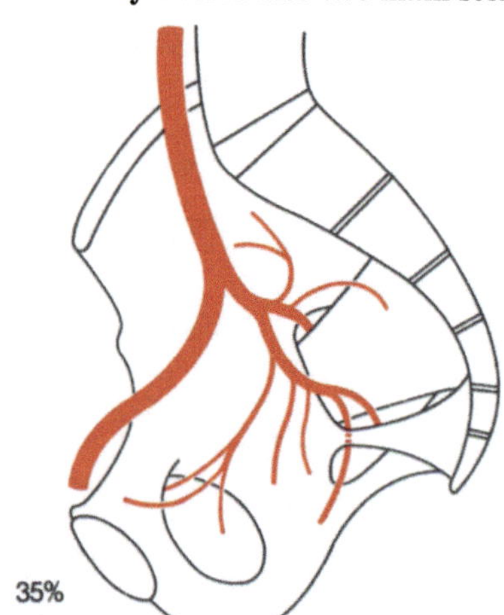

a.

a. the posterior main stem ends as the superior gluteal artery, the anterior trunk ends as the inferior gluteal and the internal pudendal arteries

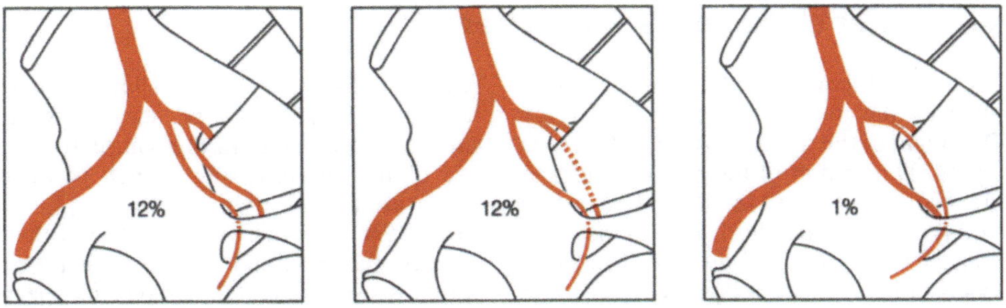

b. c. d.

2b. the anterior stem ends as the internal pudendal artery, the posterior trunk divides into the superior and inferior gluteal arteries, which leave the pelvis through the suprapiriform and infrapiriform foramen
c. same as b., but the posterior stem divides after leaving the pelvis through the suprapiriform foramen. The inferior gluteal artery then runs dorsally to the piriform muscle

d. the ventral stem ends as the inferior gluteal artery; the terminal branches of the dorsal stem are the superior gluteal and the internal pudendal arteries

3. The internal iliac artery divides into three stems which give off all other branches [20%]

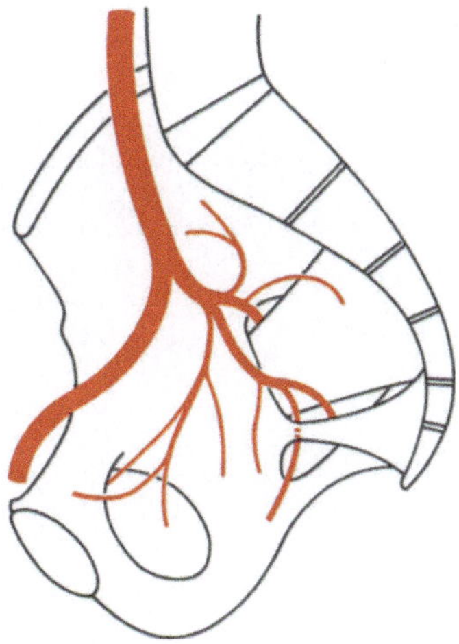

4. The internal iliac artery divides into four or more stems which branch to form the other arteries [10%]

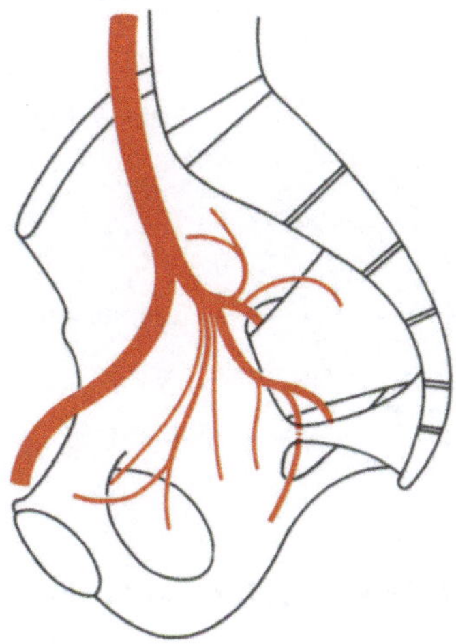

Origin of the internal iliac artery

a. the common iliac artery divides into the internal and external iliac arteries
b. the common iliac artery is absent or very short. The internal and external iliac arteries arise from the aorta

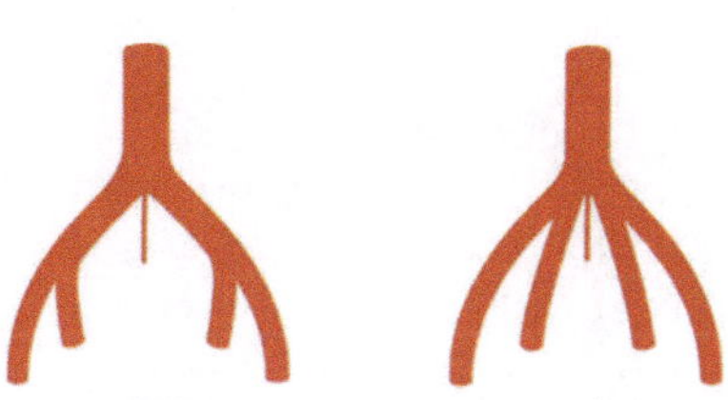

a. b.

24 Arteries of the female genital tract

The internal female genital organs are supplied by the uterine and ovarian arteries. The origin of the ovarian artery is shown on pp. 28–29 and that of the uterine artery in the chapter on the internal iliac artery (p. 54). Both arteries anastomose by branches along the uterine tube. The ovary seems to be supplied either by the ovarian artery, the ovarian branch of the uterine artery or by both. Operations on the uterine tube can disturb the blood flow in these branches. There are very few systematic studies on anomalies in the course and the size of the anastomosing branch between the ovarian and uterine arteries [3, 68, 167 a, 289 a, 304, 356], but because of their clinical relevance, and also to stimulate research on the topic, this chapter has been included.

1. "Normal" situation as described in textbooks >90%

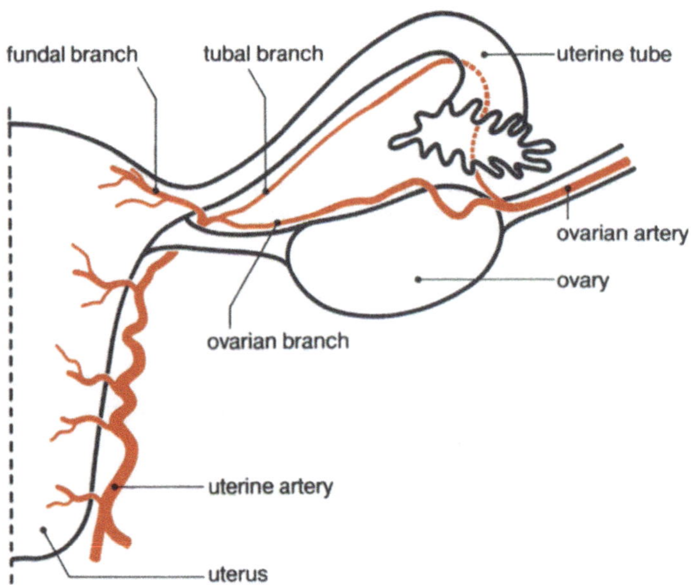

2. Blood supply of the fundus of the uterus

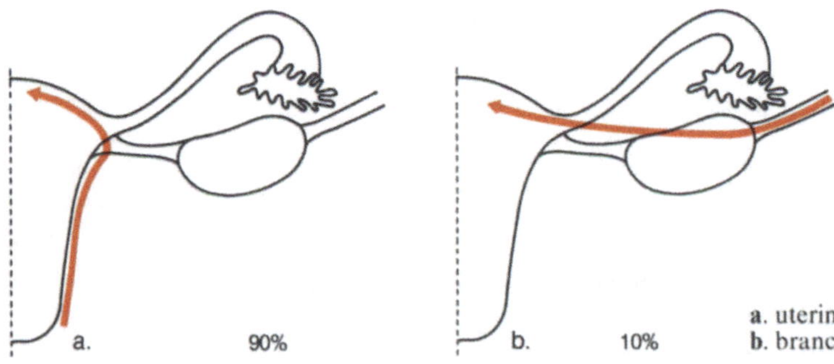

a. uterine artery
b. branches of the ovarian artery

3. Blood supply of the ovary

a.
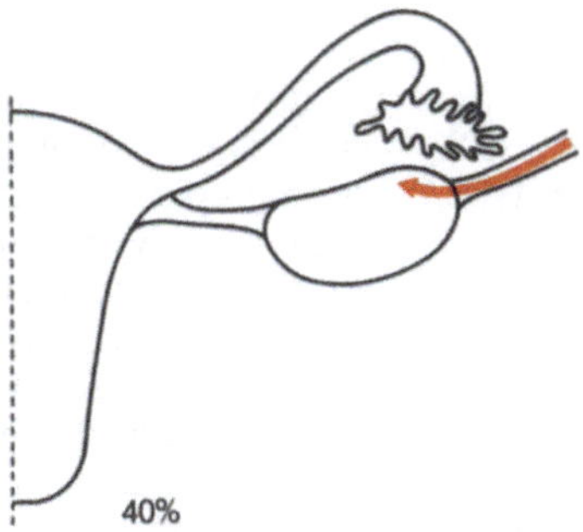

b.
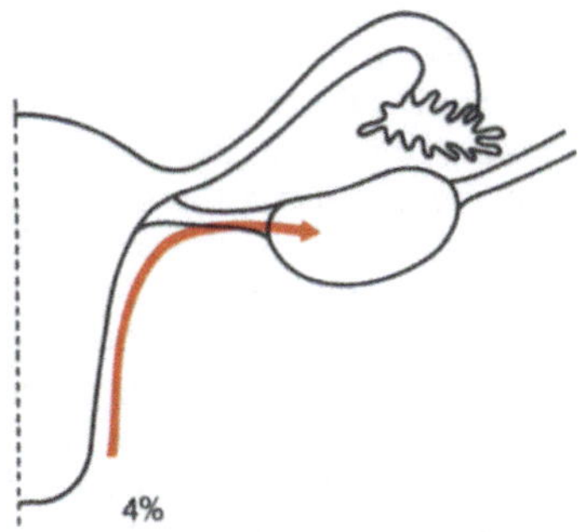

a. by the ovarian artery
b. ovarian branch of the uterine artery

c.
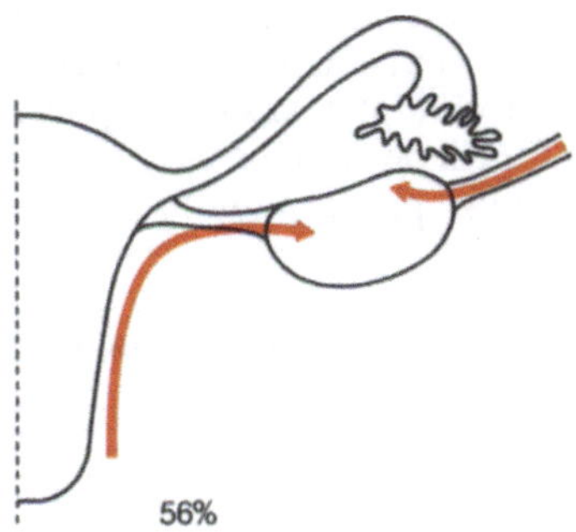

c. both the ovarian and uterine arteries supply the ovary

4. Blood supply of the uterine tube

a.
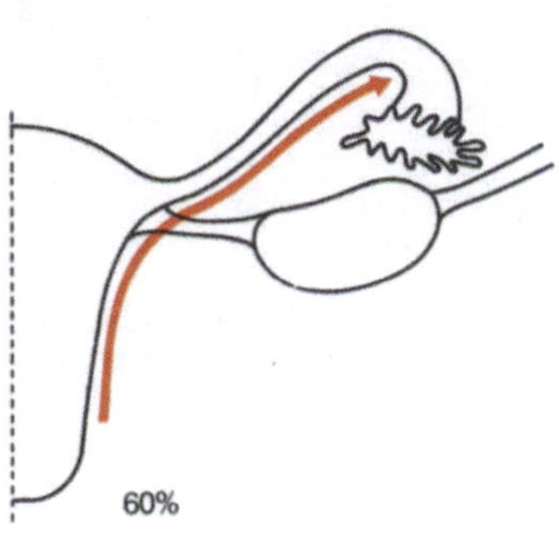

b.
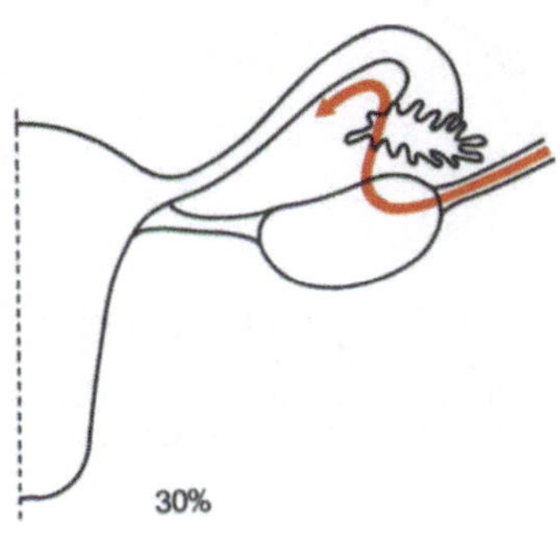

a. tubal branch of the uterine artery
b. tubal branch of the ovarian artery

c.
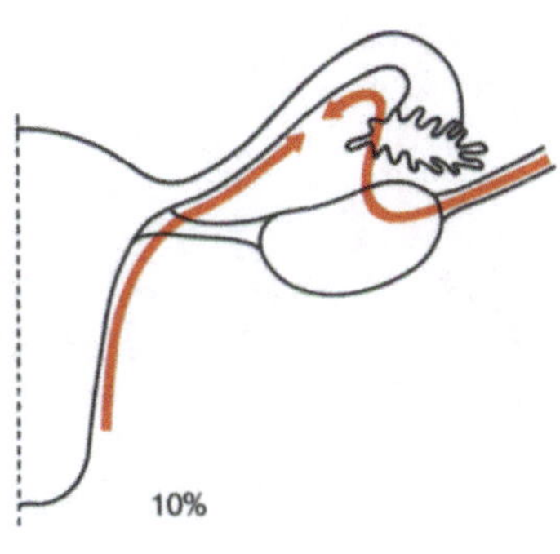

c. both the ovarian and uterine arteries supply the tube

25 Obturator artery

The number of its variations increases considerably if all the different types of branching of the internal iliac artery are considered [3*, 15, 24, 332, 356, 480*, 519]. To avoid confusion the most common type, i.e. the division of the internal iliac artery into two stems (pp. 54–55), is taken as a basis for these figures. The percentages for the different types are averages. Accessory obturator arteries are not outlined. Small accessory branches or different origins with a trunk formation are very common. In such cases arteries can cross the obturator nerve, but this has no practical relevance. The obturator artery supplies the muscles of the inner wall of the pelvis (obturator internus, psoas muscle) and has a pubic branch to the symphysis, which anastomoses with the obturator branch of the inferior epigastric artery. In Fig. 2c this anastomosis is very prominent and in Fig. 2b the inferior epigastric artery alone feeds the obturator artery. A major branch between the inferior epigastric artery and the obturator artery was formerly called the "a. corona mortis". This artery was sometimes lacerated during herniotomies and because such arterial bleeding was difficult to staunch in former times this type of anomaly could prove fatal. After passing the obturator canal, the obturator artery divides into the anterior branch to the adductor muscles and the posterior branch to the hip joint. A small acetabular branch reaches the head

1. Origin from the internal iliac artery 75%

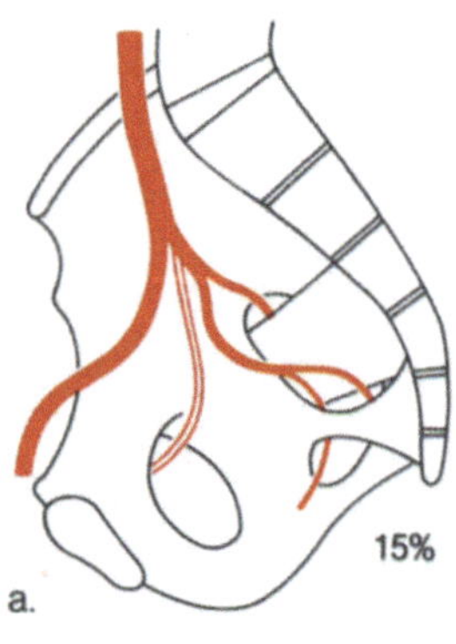

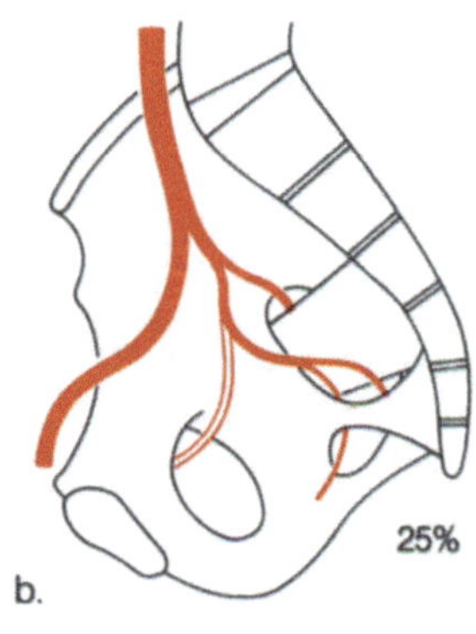

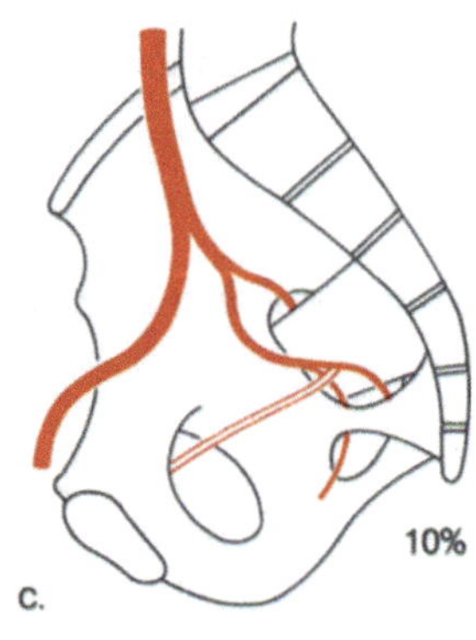

a. the obturator artery branches from the stem of the internal iliac artery
b. the obturator artery branches from the anterior stem of the internal iliac artery

c. the obturator artery branches from the inferior gluteal artery

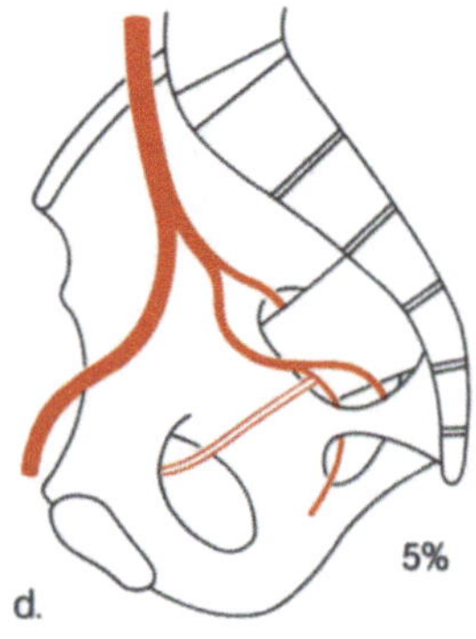

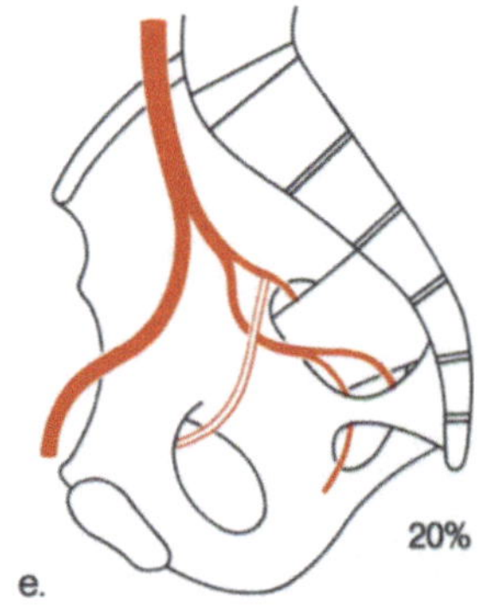

d. the obturator artery branches from the internal pudendal artery
e. the obturator artery branches from the dorsal stem of the superior gluteal artery

of the femur through the ligament of the head of the femur. Normally this branch is not able to supply sufficient blood to the head of the femur in cases of an epiphysiolysis. Racial differences have been discovered in connection with the origin of the obturator artery from the external iliac artery. The data given here are derived from studies on North Americans and Europeans; however, in Japan the frequency of this anomaly seems to be halved [3]. In females this anomaly is assumed to be more frequent. The obturator artery can also originate from the femoral artery; it then turns backwards through the lacuna vasorum into the pelvis [332]. The anastomoses between the obturator artery and the inferior epigastric and other arteries of the leg are sometimes of importance when the external iliac artery is occluded in atherosclerosis.

2. Origin from the external iliac artery 25%

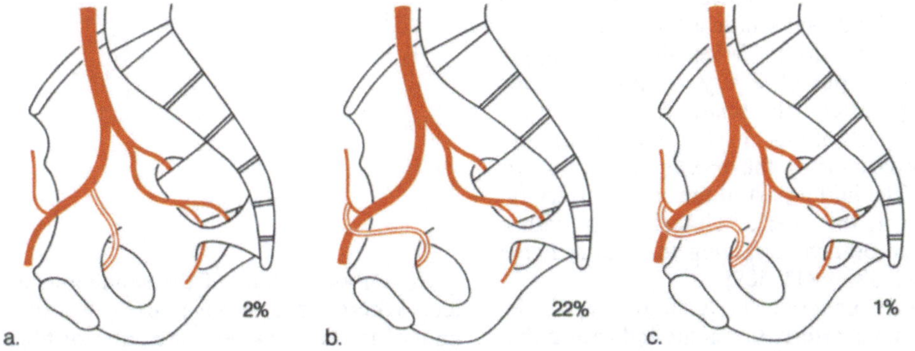

a. the obturator artery branches directly from the external iliac artery
b. the obturator artery is a branch of the inferior epigastric artery
c. the obturator artery originates with roots of about equal size from the external iliac and the internal iliac or its branches

Accessory pudendal artery

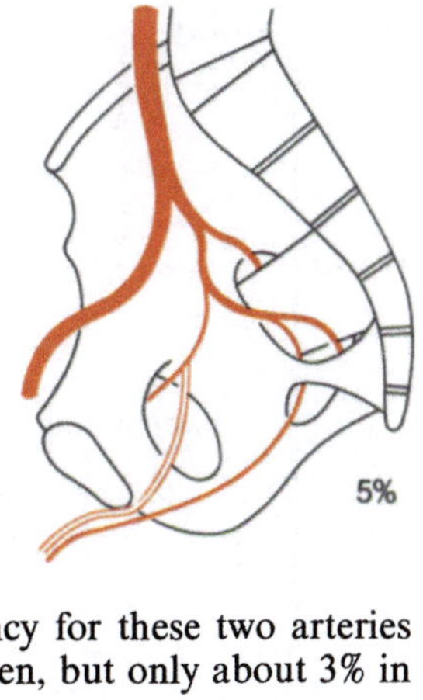

The obturator artery usually gives off tiny branches to the area of the internal pudendal artery. In about 5% of all cases, these branches take over part of the blood supply of the internal pudendal artery and end as the dorsal or deep artery of the penis or clitoris [3]. Such a branch is then called an accessory pudendal artery. In another 5% the accessory pudendal artery arises directly from the internal iliac artery or the initial part of the internal pudendal artery [3, 617]. Always present in some lower mammals, this vessel is called the urogenital artery, since it also supplies the bladder [234]. The frequency for these two arteries is about 10% in men, but only about 3% in women.

26 Development of the arteries of the lower limb

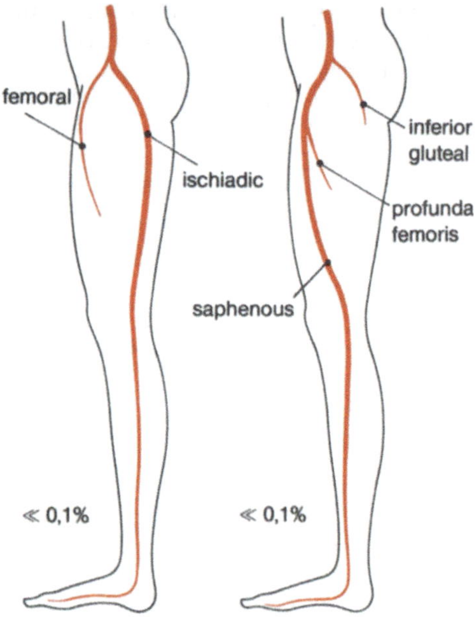

The arteries of the lower limbs derive from the 5th lumbar artery (p. 23) which forms the umbilical artery in the embryo. At the embryo stage of crown-to-heal length of 10 mm, several side branches can already be identified: external iliac, sciatic, superior gluteal and internal pudendal arteries. The sciatic artery passes through the sciatic plexus and forms the main artery of the leg. It persists in most vertebrates. By contrast, in mammals the femoral artery, as the continuation of the external iliac artery, becomes the main artery for the lower limbs. Very early, anastomoses are formed between the dorsal sciatic artery and the ventral femoral artery. When the main supply to the popliteal artery comes from the femoral artery, the sciatic artery regresses. Other parts of this anastomotic network are the precursor or the profunda femoral artery. The initial part of the sciatic artery remains as the inferior gluteal artery with a minute branch supplying the sciatic nerve [3, 213, 330, 484*, 589].

Another branch of the femoral artery is the saphenous artery, the name indicating that if follows the same course as the saphenous nerve and the great saphenous vein. This artery is connected to the arteries of the foot; its connections to the sciatic artery disappear. As some parts atrophy while others grow, the descending genicular artery and the posterior tibial artery are finally the only remnants of the saphenous artery. Whereas the peroneal artery is a remnant of the sciatic artery, the anterior tibial artery is a new branch. This complicated pattern of embryological development in the arteries of the leg explains the large number of anomalies [3, 213, 293, 373, 484, 555].

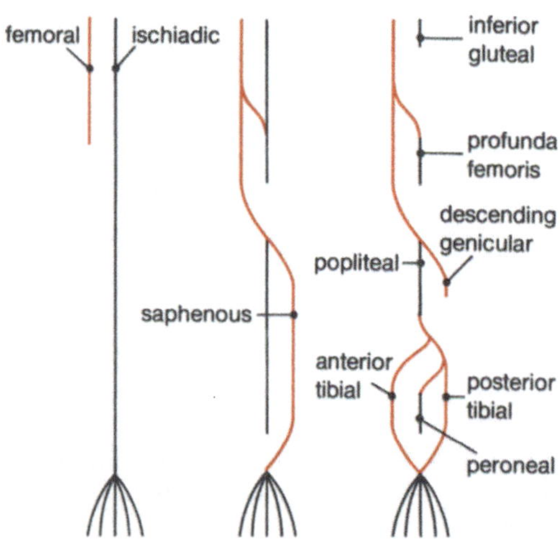

Persistent sciatic artery

Although a rare anomaly in humans recently many case reports have been published on the persistent sciatic artery [72, 132, 171, 172, 385, 391, 484, 488, 662, 675]. The sciatic artery, as a branch of the internal iliac artery, leaves the pelvis via the infrapiriform foramen and runs parallel to the sciatic nerve to form the popliteal artery. In such cases the femoral artery is small and only supplies the area of the a. profunda femoris.

Persistent saphenous artery

This artery is regularly found in many monkeys but surprisingly seldom in humans. It is a superficial artery running along the great saphenous vein [330].

27 The profunda femoris artery

[3, 37, 56, 94, 213, 214, 308, 330, 373, 423, 473, 485, 499]

1. Origin and branching pattern

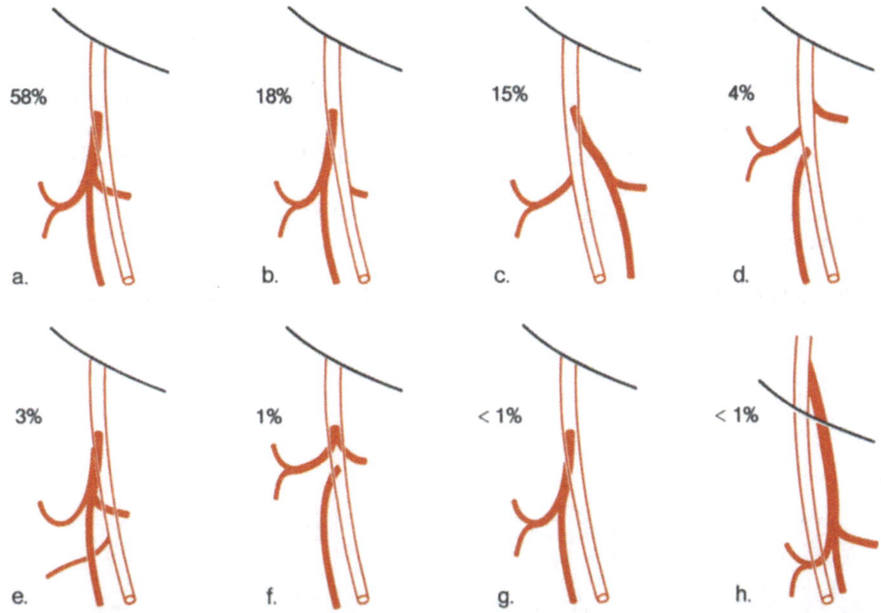

a. a common trunk supplies the profunda femoris, the medial and lateral circumflex arteries of the femur
b. the medial circumflex artery is a direct branch of the femoral artery
c. the lateral circumflex artery is a direct branch of the femoral artery
d. both circumflex arteries originate directly from the femoral artery
e. same as a. or b., but a major branch of the lateral circumflex artery comes from the femoral artery
f. both circumflex arteries have a common trunk independent of the profunda femoris
g. there is no medial circumflex artery; the area is supplied by the obturatorial artery
h. the profunda femoris is a branch of the external iliac, sometimes together with the inferior epigastric artery

2. Course

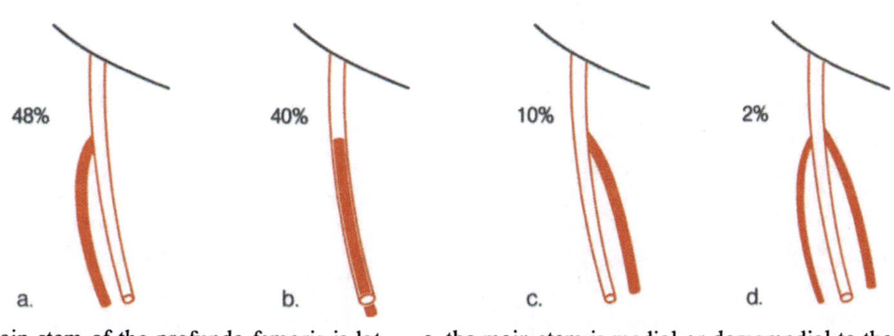

a. the main stem of the profunda femoris is lateral or dorsolateral to the femoral artery
b. the main stem is dorsal to the femoral artery
c. the main stem is medial or dorsomedial to the femoral artery
d. large branches of the profunda femoris are found medial as well as lateral to the femoral artery

28 Popliteal artery

Normally the popliteal artery divides at the upper border of the popliteal muscle. If the division is proximal to this border, it is called a high division. The anterior tibial artery runs ventrally to the popliteal muscle and corresponds to the profound popliteal artery of most mammals. Only in primates is this vessel replaced by the more superficially positioned popliteal artery. An interosseous artery is very rarely found along the interosseous membrane; this artery differs from the popliteal artery [330, 397, 423, 557, 620].

In addition to anomalies of the main branches of the popliteal artery, there are many more varieties of the smaller side branches (the genicular and sural arteries). For example, common origins of the superior medial genicular and the descending genicular arteries as well as the superior lateral and the median genicular arteries are frequent. Both upper and lower arteries of the knee joint only rarely form trunks. The origin of the genicular arteries from the sural arteries has been described.

1. Branching of the popliteal artery at the lower border of the popliteal muscle 95%

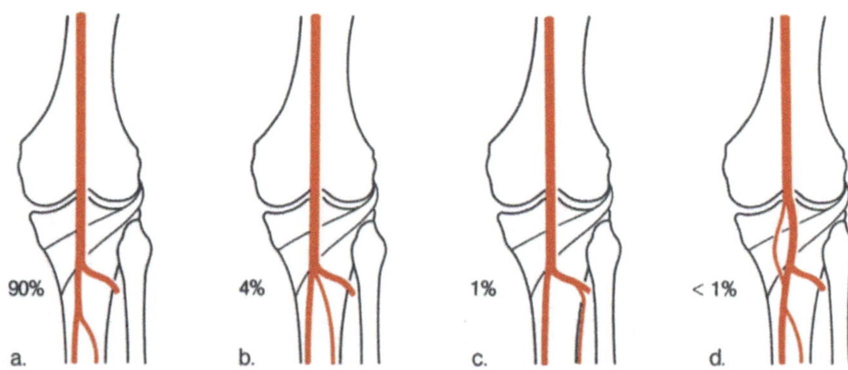

a. the "normal" situation as described in textbooks
b. "trifurcation" or very short stems of the anterior, posterior tibial and peroneal arteries
c. trunk to the anterior tibial and peroneal arteries
d. "island" formation of a branch

2. Branching of the popliteal artery proximal to the upper margin of the popliteal muscle 5%

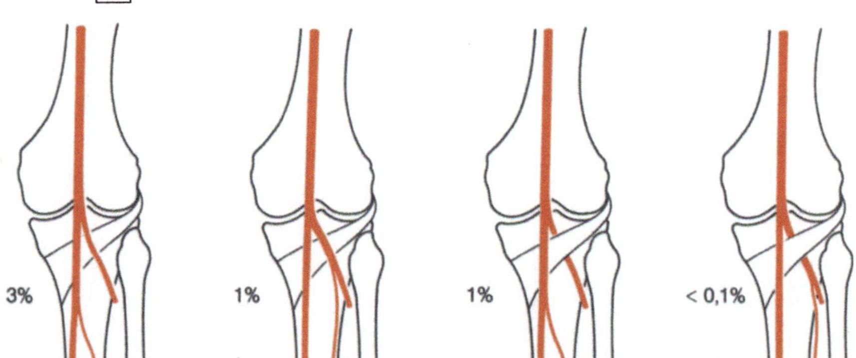

a. otherwise as normal
b. combined with a trunk formation (Tr. peroneotibialis anterior)
c. the anterior tibial artery runs ventrally to the popliteal muscle
d. trunk formation: interosseous and anterior tibial artery

29 Arteries of the lower leg

The peroneal artery is the most consistently encountered artery in the lower leg. Parts of this artery stem from the original artery of the lower limb (p. 60). The variations in the arteries of the lower leg can be best explained by treating the peroneal artery as the main artery and defining the tibial arteries as its branches. If one of the tibial arteries is lacking or very small, the peroneal artery supplies that part of the foot. If the anterior tibial artery is absent, the perforans branch of the peroneal artery forms the dorsal artery of the foot, or if the posterior tibial artery is absent, the peroneal artery forms the plantar arteries. The peroneal artery is therefore a major contributor to the blood supply of the foot in about 12% of all cases. Normally, however, the peroneal artery ends in the malleolar and the calcanean areas [3, 330, 373, 482].

An interesting variability has been described for the nutrient artery of the tibia. In about 50% of all cases this artery branches from the posterior peroneotibial trunk, or from the anterior tibial or posterior tibial arteries, or more rarely from the peroneal or popliteal arteries. Depending on the origin of the nutrient artery, the nutrient foramen is found on the medial (a. tibialis posterior) or on the lateral aspect (a. peronea, a. tibialis anterior).

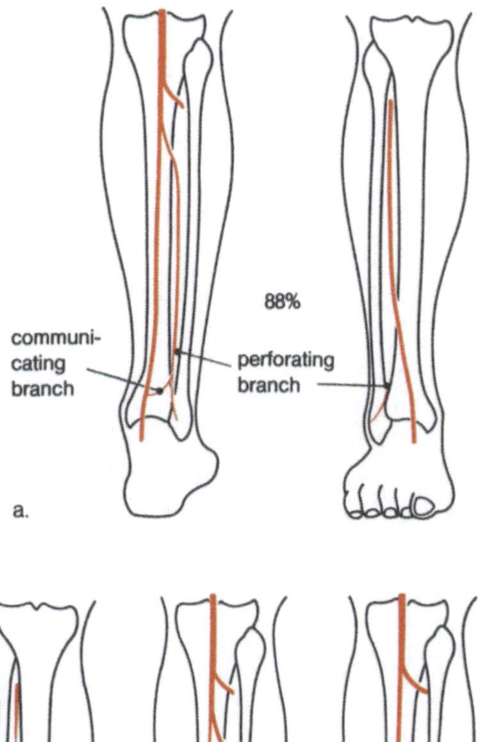

a. the normal situation as shown in textbooks: the peroneal artery has a communicating branch to the posterior tibial artery and a perforating branch to the anterior tibial artery

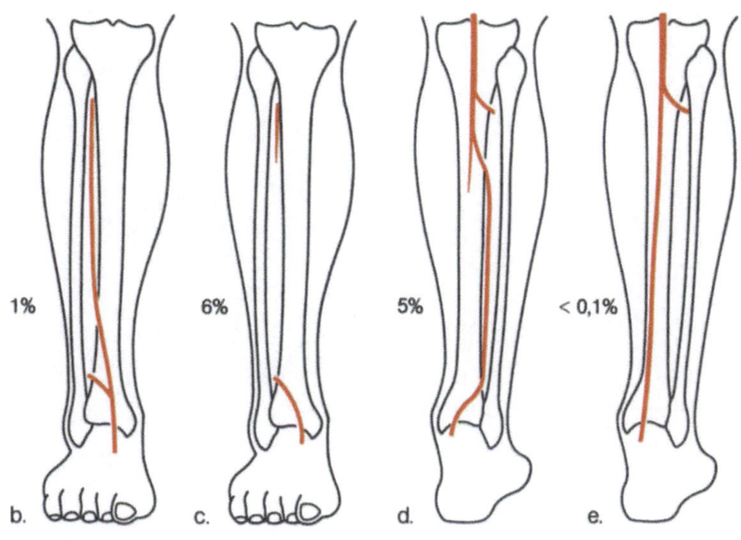

b. the dorsal pedis artery originates from two roots of equal size from the peroneal and anterior tibial arteries
c. the dorsal pedis artery originates from the perforating branch of the peroneal artery; the anterior tibial artery is absent or ends at the lower leg
d. the posterior tibial artery is absent or ends at the lower leg; the plantar arteries branch from the peroneal artery
e. the peroneal artery is absent

30 Dorsal arteries of the foot

The anterior tibial artery ends as the dorsal artery of the foot, which branches into the lateral tarsal artery and the medial tarsal arteries and ends as the arcuate artery. This forms the dorsal metatarsal arteries which continue as the dorsal digital arteries. In the first interosseous space the prominent deep plantar branch is connected with the plantar arch. The other dorsal metatarsal arteries are also connected with the plantar arteries. These perforating branches can replace the dorsal metatarsal arteries. The great variety of combinations of a more dorsal (D) or plantar (P) origin of the arteries on the dorsal side of the foot is shown. The figures do not include all the possibilities, but those omitted account for only 3–4% of all cases [3, 143, 267, 330, 482].

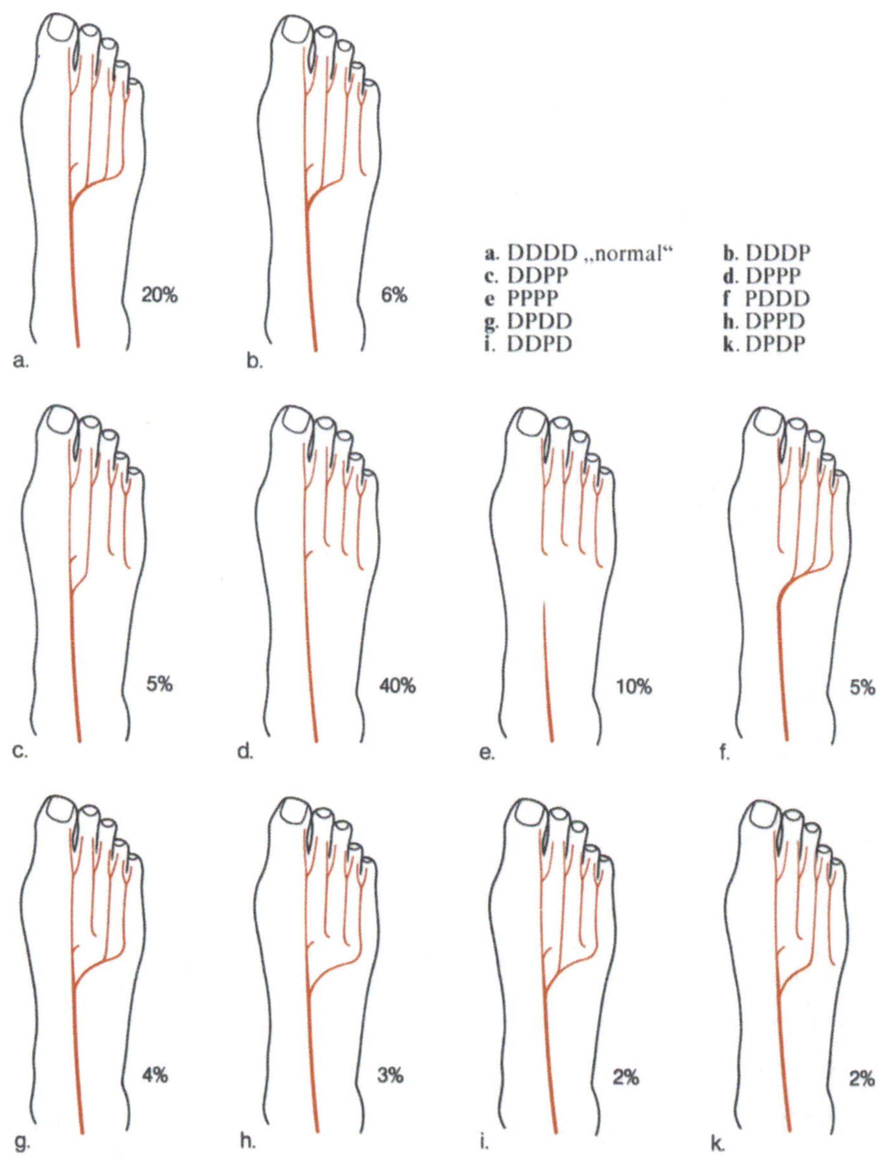

31 Plantar arch

The plantar arch is formed by the lateral plantar artery and the deep plantar branch of the dorsalis pedis artery. In only 1% of all cases (type h) is there no connection between these two arteries. Normally, one artery is obviously the feeding vessel, and the areas supplied by the plantar (P) or dorsal (D) arteries can be distinguished. The plantar arch is comparable to the deep palmar arch, in which the major supply also comes from the dorsal side (radial artery). A superficial arch as occurs in the palm is rarely found in the planta pedis: a major superficial plantar arch occurs in about 2% and small anastomoses in about 25% of all cases [3, 143, 330, 482].

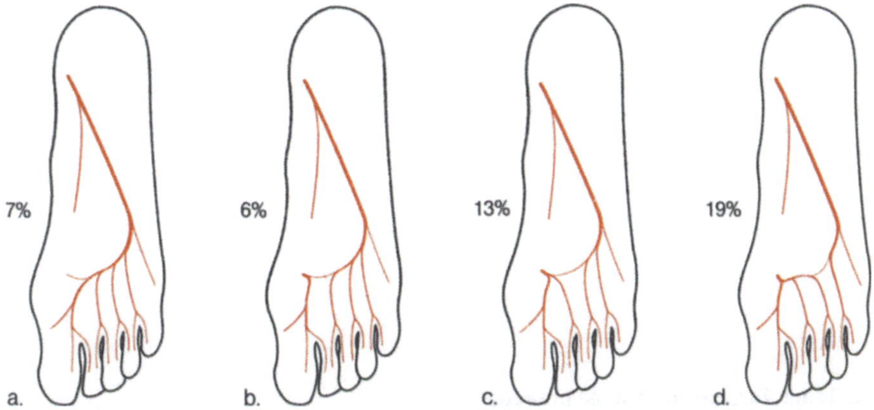

a. all four plantar metatarsal and common digital arteries respectively are derived from the lateral plantar artery
b. the first plantar metatarsal artery derives from the deep plantar branch
c. the first and second plantar metatarsal arteries derive from the deep plantar branch
d. the plantar metatarsal arteries I–III derive from the deep plantar branch

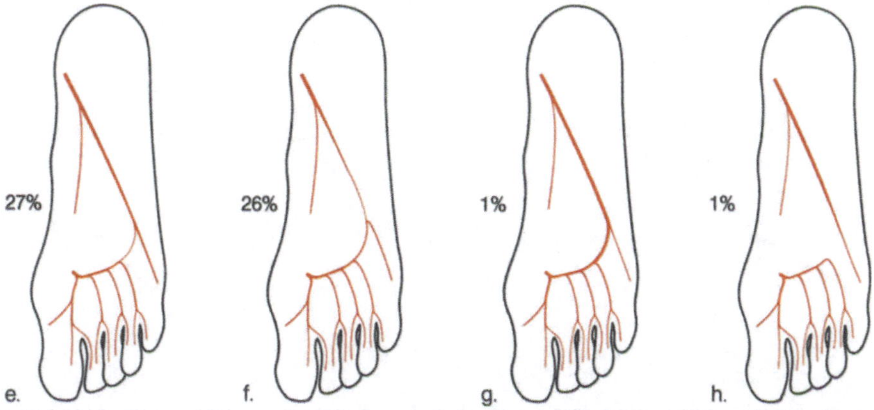

e. all four plantar metatarsal arteries derive from the deep plantar branch
f. the deep branch also reaches the lateral part of the foot
g. in all areas an equally strong plantar arch without an obvious boundary between the areas supplied by the lateral plantar artery and the deep plantar branch
h. the plantar arch is not closed

32 Axillary artery

Enormous differences exist in the varieties of the axillary artery. Racial differences seem to play a role, but more important are the variations in the trunks' names. The percentages given should be taken as rough estimates. Type 2 has an overall frequency of about 50% less than the sum of all subtypes 2a–d, because the combination 2e has been included. More exact data are not available [3, 15*, 95, 107, 190, 269, 293, 332*, 405, 412, 576, 580, 621, 646].

1. Normal type as shown in textbooks < 10%?

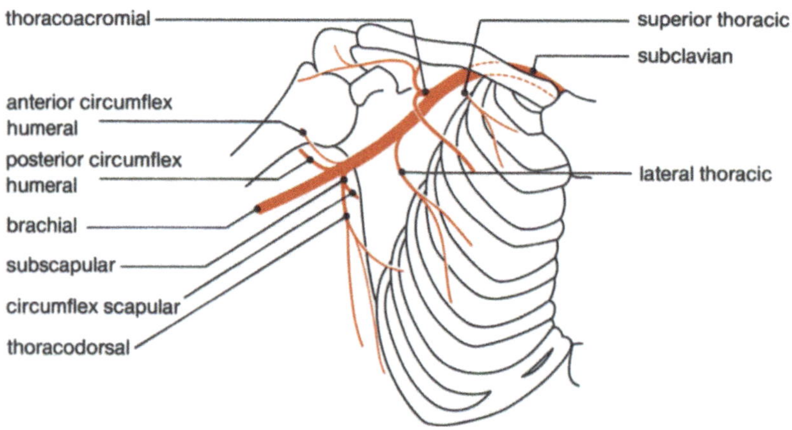

2. Trunk formation of some branches ~ 50%

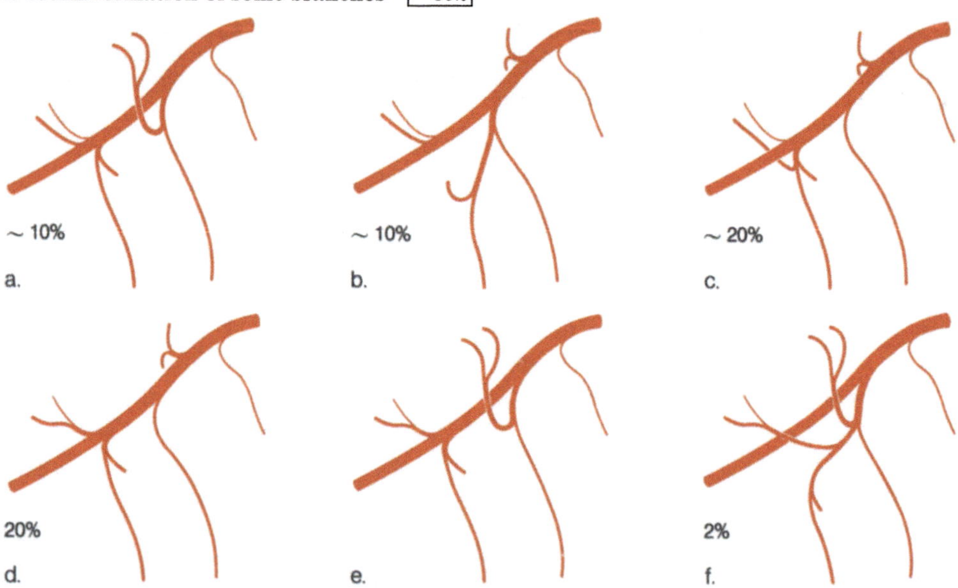

a. the lateral thoracic artery forms a trunk with the thoracodorsal artery
b. the lateral thoracic artery originates with the subscapular artery
c. the subscapular artery forms a trunk with the posterior circumflex artery of the humerus
d. both circumflex arteries of the humerus form a trunk
e. combination of types 2a–d (e.g. 2a+2d)
f. the majority of branches originate as a common trunk from the axillary artery

3. High origin of arteries of the arm 22%

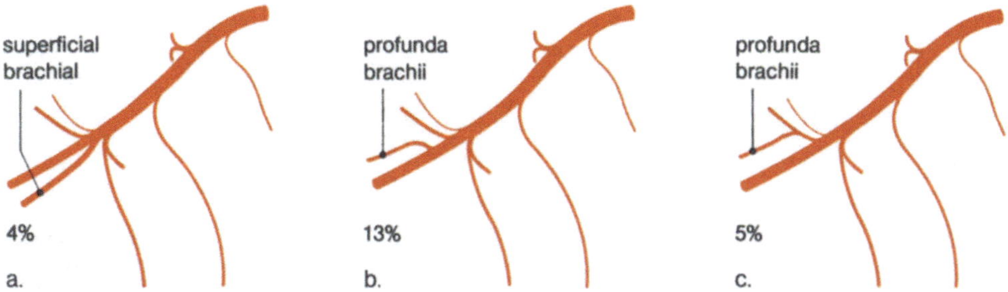

a. a superficial brachial artery derives from the axillary artery
b. the deep brachial artery originates from the axillary artery
c. the deep brachial artery is a branch of the posterior circumflex artery of the humerus

4. Other variations 21%

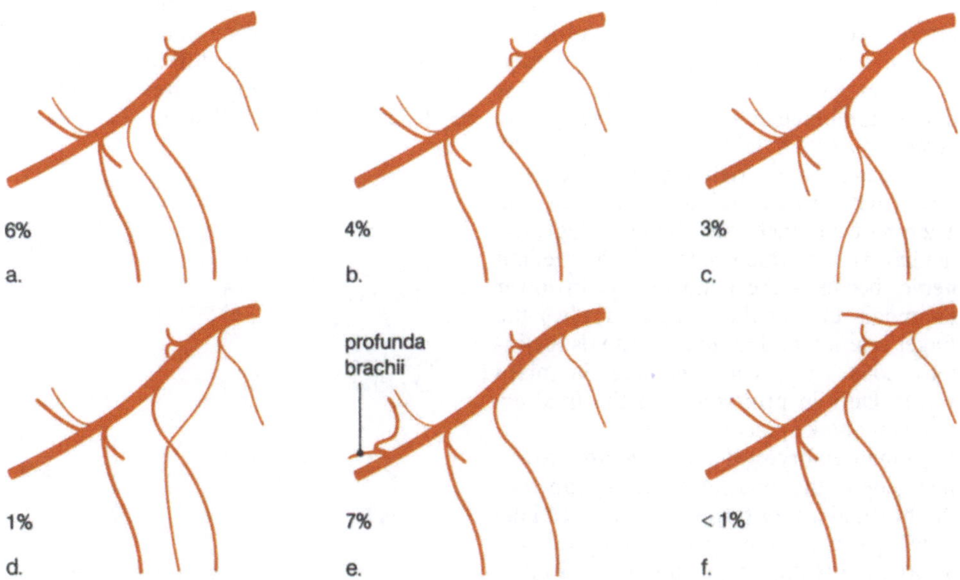

a. accessory branches to the thoracic wall
b. the scapular circumflex artery and thoracodorsal arteries originate separately from the axillary artery
c. the lateral thoracic artery is prominent, the thoracodorsal artery small
d. the superior thoracic artery has a long descending branch
e. the posterior circumflex artery of the humerus originates from the brachial or deep brachial artery
f. the suprascapular artery is a branch of the axillary artery instead of the thyrocervical trunk

33 Development of the arteries of the arm

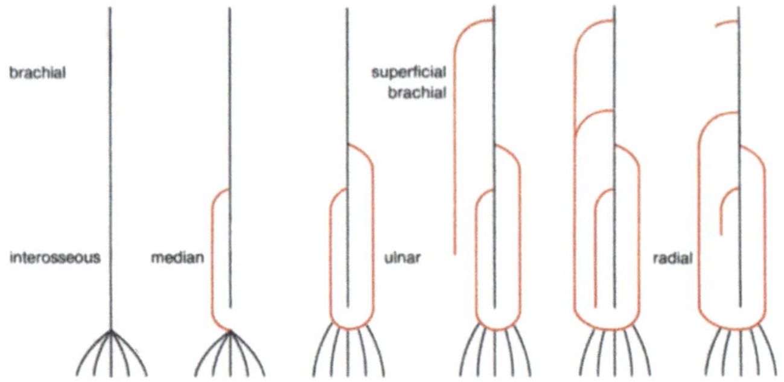

A very schematic drawing of the development of the arteries of the upper extremity: the original system in black, arteries formed later in red

The arteries of the arm show many rearrangements, especially the arteries of the forearm which develop like the arteries in the leg. The axial artery of the arm is the continuation of the 6th cervical segmental artery (see p. 3). This artery has different names: it starts as the subclavian artery, continues as the axillary artery and is named the brachial artery after passing the posterior axillary fold. In the forearm the continuation was originally the interosseous artery, which initially supplied the arteries of the fingers. At the next stage, the median artery, which runs with the median nerve, becomes the main artery. In lower mammals the median artery supplies the finger arteries, and in non-mammals the interosseous artery even remains the main vessel. Only in primates does the final arrangement take place.
In human embryos of about 18 mm crown-heel length, the ulnar artery as a branch of the brachial artery forms the superficial arterial palmar arch, which supplies the arteries of the fingers. In embryos of 21 mm length the superficial vessels widen, starting from the axillary region. This superficial brachial artery turns to the lateral side and ends on the dorsal side of the hand. This vessel also connects to the palmar arch and the median artery regresses (embryo of 23 mm length). An anastomosis between the superficial brachial and the brachial artery increases in size, while the proximal part of the superficial brachial artery normally disappears. What remains is the radial artery. This complicated development explains the many variants observed. However, since anastomoses are formed with other additional superficial arteries of the forearm (superficial antebrachial artery), the picture becomes even more confusing. The schematic drawing shows the general possibilities, with the different types outlined in the following figures [3, 239, 317, 332, 555*, 574, 589, 629, 650].

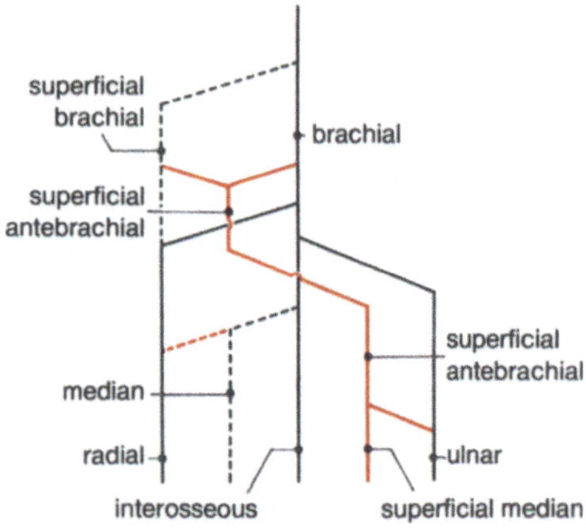

Variants of the arteries of the forearm:

- —— "normally" persisting vessels
- ----- "normally" regressing vessels
- —— system of superficial antebrachial artery
- ----- anastomosis between the radial and median artery

34 Brachial artery and superficial brachial artery

The superficial brachial artery is the name of a major artery which is superficial to the median nerve on the upper arm. Such an artery can replace the normal brachial artery or occur in addition to it. To calculate the frequency, only arteries were included which reach the forearm, not minor branches such as those to the biceps muscle. The superficial brachial artery normally branches from the proximal part of the upper arm and only rarely from the distal half [3, 15*, 239, 283, 332, 361, 377, 387*, 460, 550, 555, 575, 576, 580, 649].

The practical importance of this artery is that it often runs to the forearm in front of the aponeurosis of the biceps muscle and can be mistaken for a vein. An "intravenous" injection in such an artery can result in the loss of the hand or forearm. One should keep in mind that every fifth patient has a superficial artery in the cubital fossa which could be mistaken for a vein [251, 474, 486].

1. Only a brachial artery 78%

a. "normal" situation as shown in textbooks: the artery is positioned behind the median nerve, crossing beneath it at the upper arm
b. there is no typical median nerve in the axilla; the artery crosses below the nerve as usual

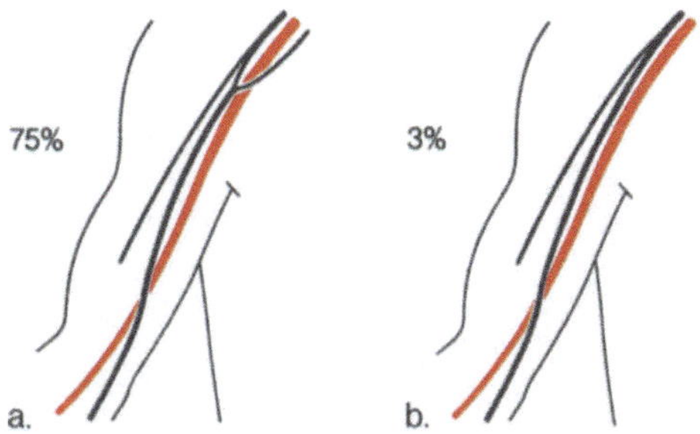

2. Only a superficial brachial artery 9%

a. main artery in front of the two roots of the median nerve
b. main artery behind the two roots of the median nerve, but crossing below it at the upper arm

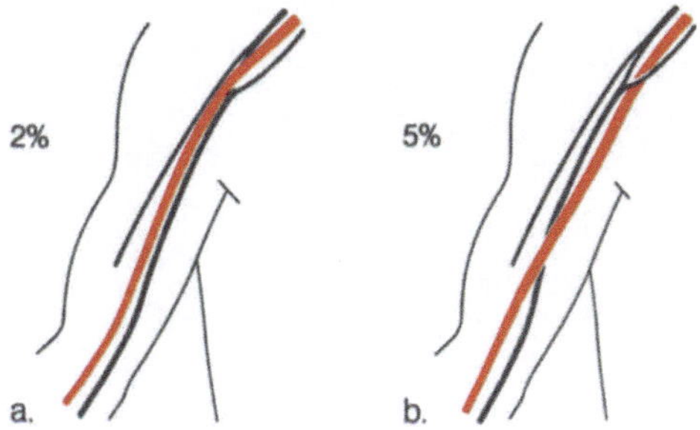

c. the artery crosses beneath the median nerve on the dorsal side, but over the nerve branch between the musculocutaneous and median nerve
d. no typical formation of the median nerve in two roots; the artery crosses over the median nerve

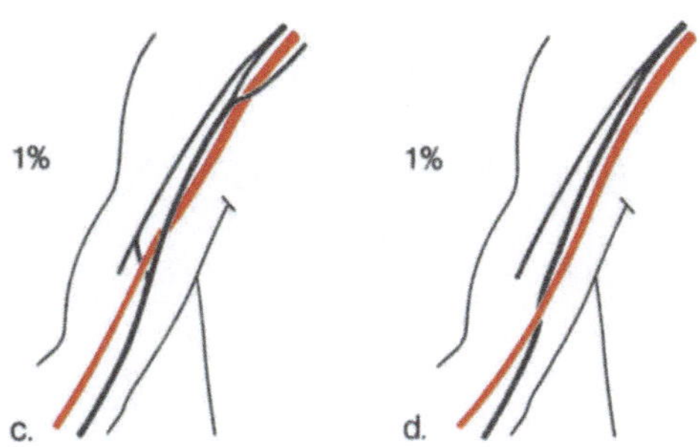

3. Two main arterial stems 13%

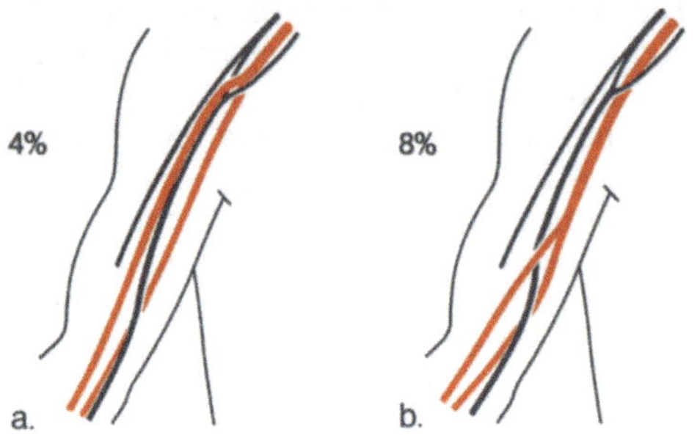

a. the axillary artery divides into two stems, one running behind, the other in front of the two roots of the median nerve
b. the brachial artery divides into two stems, one artery running in front of the median nerve

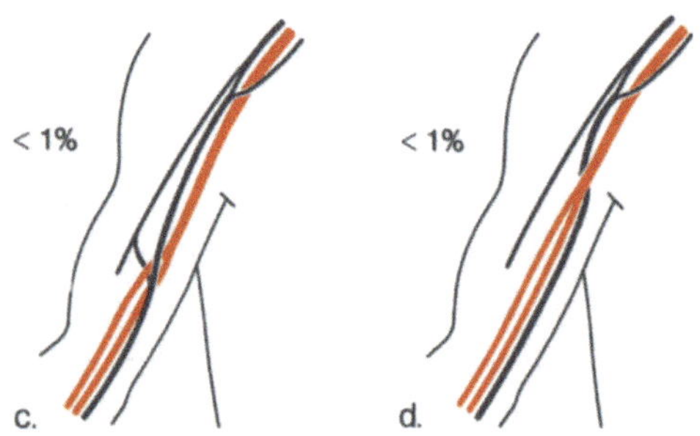

c. the brachial artery divides into two stems; one artery crosses over the nerve anastomosis between the musculocutaneous and median nerve
d. high division of a superficial brachial artery (both running superficially)

35 Arteries of the forearm

Type 1 summarizes all cases in which the arteries of the forearm follow a "normal" course, leaving the origin of these arteries out of consideration. The schematic drawings demonstrate the embryological basis for the more important variations. The origin of an ulnar artery as a branch of the superficial brachial artery running parallel to a brachial artery (p. 70, type 3) has not been described. The superficial arteries of the forearm usually branch from a superficial brachial artery, which runs parallel to the brachial artery (p. 70, type 3). Sometimes the superficial arteries of the forearm are branches of the brachial artery [3, 15, 283, 332, 361, 387*, 460, 627, 653, 580].

1. "Normal" arterial pattern of the forearm (radial, ulnar, and interosseous artery) 84%

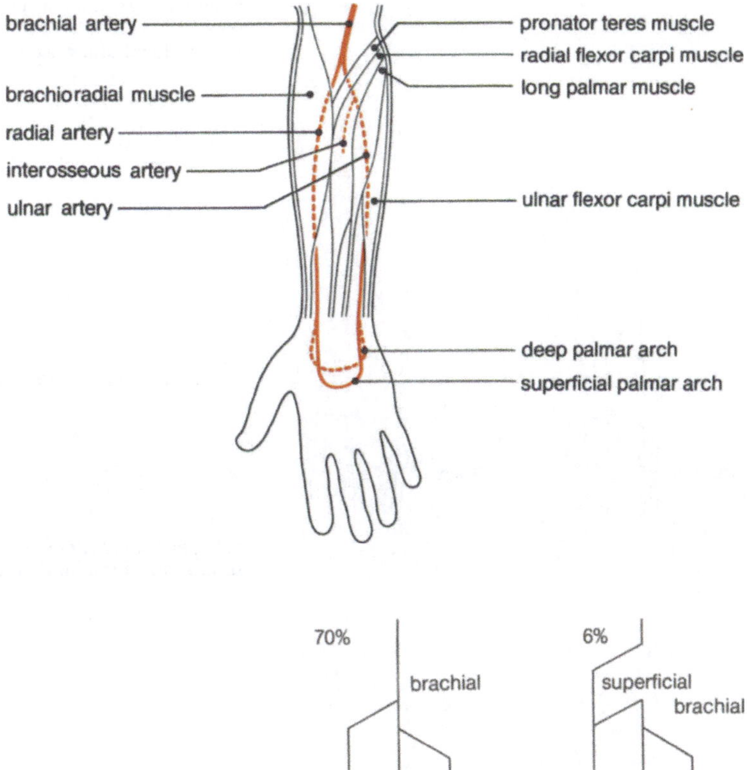

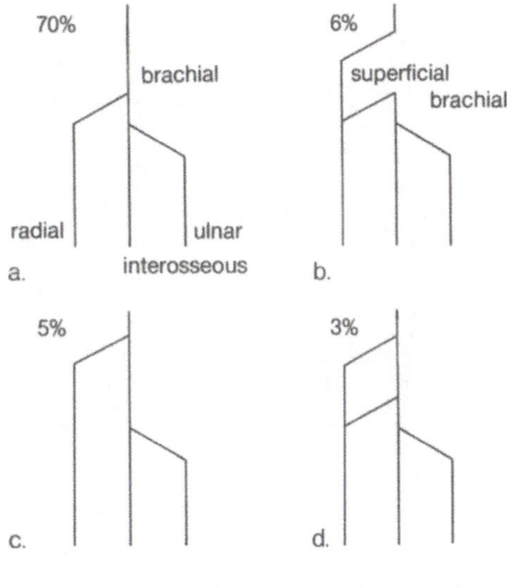

a. all arteries of the forearm branch from the brachial artery (p. 69, type 1)
b. all arteries of the forearm are branches of a superficial brachial artery

c. the radial artery originates from the superficial brachial artery, the ulnar and interosseous artery from the brachial artery
d. same as c., but with a large anastomosis between the brachial and radial artery in the cubital region

2. Superficial arteries of the forearm (a. antebrachialis superficialis) 8%

3%

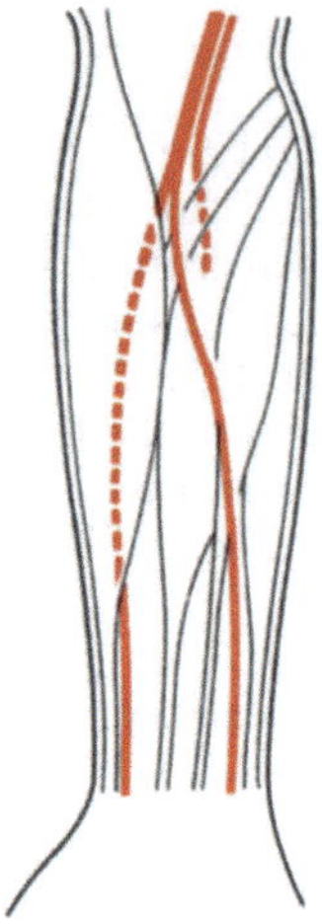

a.

< 1%

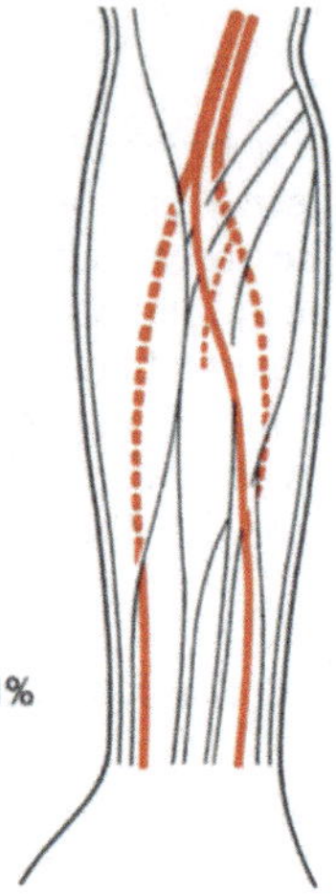

b.

a. superficial ulnar artery replaces the ulnar artery
b. superficial ulnar artery joins the ulnar artery

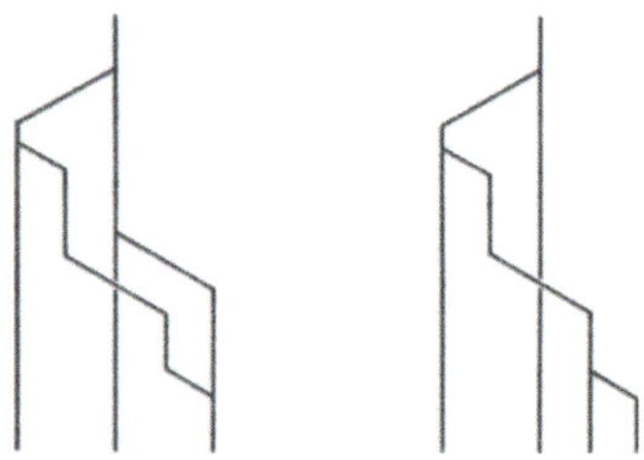

Schematic drawings to the types 2 b and 2 d

2%

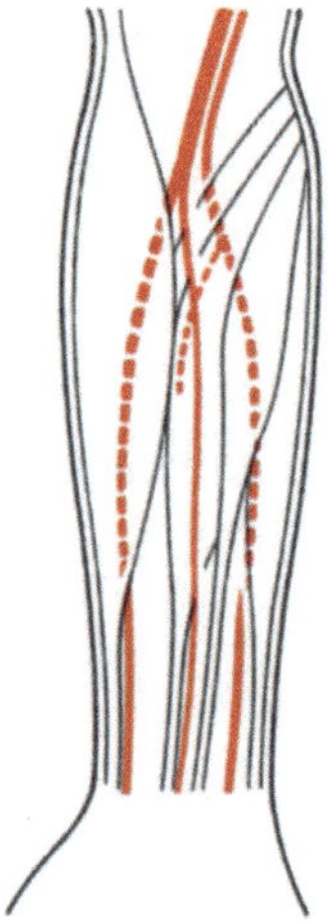

c.

< 1%

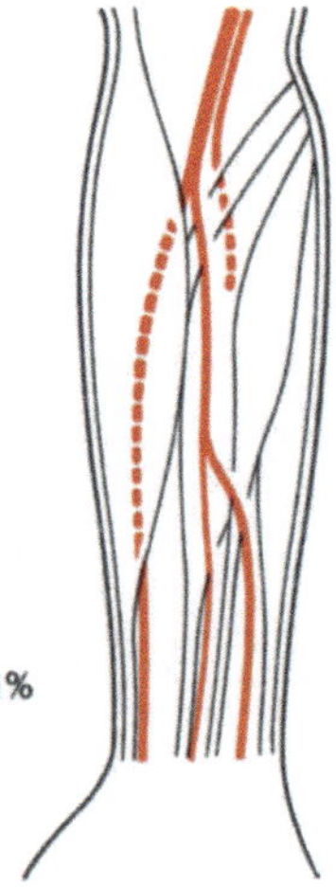

d.

c. superficial median artery
d. superficial medioulnar artery

1%

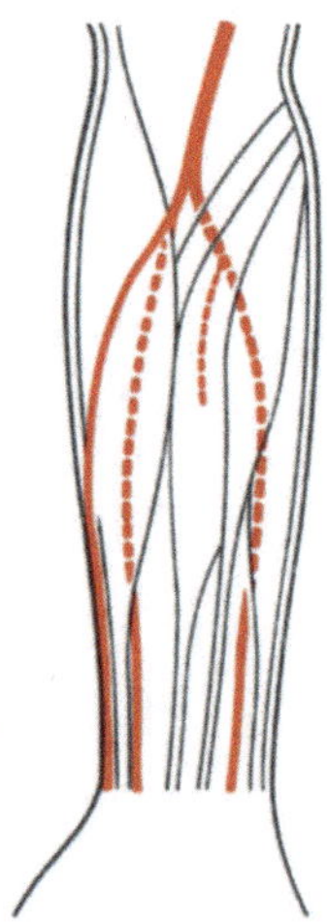

e.

1%

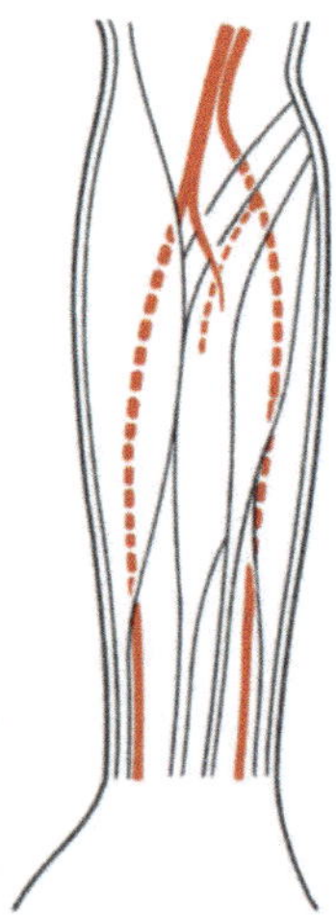

f.

e. superficial radial artery in addition to a normal radial artery
f. the superficial antebrachial artery ends at the proximal part of the forearm

3. Median artery 8%

The median artery is present in early embryonic development, but regresses as early as the second embryonic month (p. 68). This regression can occur much later, resulting in the obliteration of parts of the median artery in adults. Normally the median artery runs parallel to the median nerve, but superficial median arteries have also been described (type 2c, d; p. 69), which pass the flexor muscles obliquely and reach the median nerve further distally [3, 15, 349, 460].

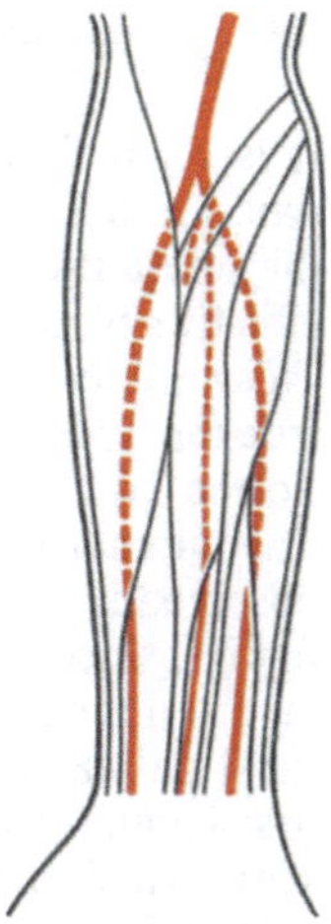

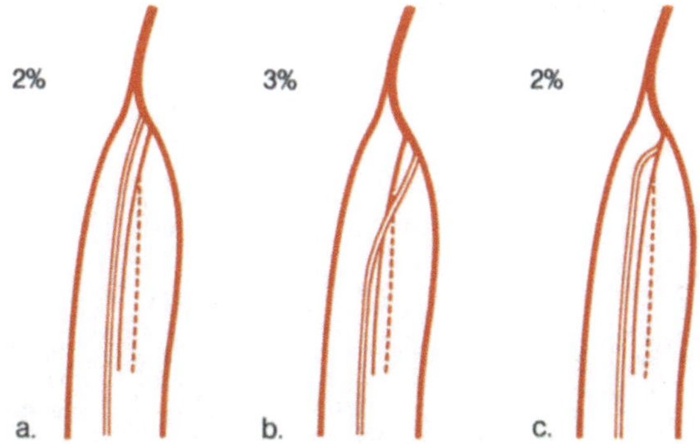

a. origin from the ulnar artery proximal to or with the common interosseous artery
b. origin from the ulnar artery distal to the common interosseous artery
c. origin from the common interosseous artery

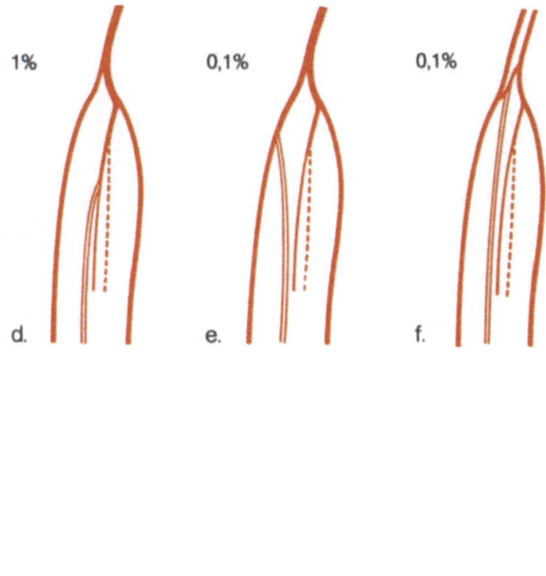

d. origin from the anterior interosseous artery
e. origin from the radial artery
f. origin from an anastomosis between the radial and brachial artery

36 Superficial palmar arch

It is often difficult to distinguish between type 1 and 2 of the superficial palmar arch: the anastomosis can be so minute that an effective blood flow can hardly take place. This problem in classification is especially true for types 1 a and 2 a–c, 1 b and 2 e, and 1 c and 2 d.

The number of types of anomalies actually observed is even larger than shown in these figures, since all four common palmar digital arteries do not originate from the superficial palmar arch (p. 76). Quite often the first common palmar digital artery is absent, which normally supplies the ulnar side of the thumb and the radial side of the index finger. In such cases the princeps pollicis artery is the only artery for that region. Because of the frequent absence of the first common palmar digital artery only the next artery is sometimes designated the common palmar digital artery. This is called II in our figures because it is found in the second interosseous space.

In these figures no differentiation has been made as to whether the feeding median artery is a "normal" median artery or a superficial median artery in the forearm [3, 11, 15, 103*, 125, 136, 144, 288, 332, 350, 359, 422, 460, 636].

1. Closed arch 42%

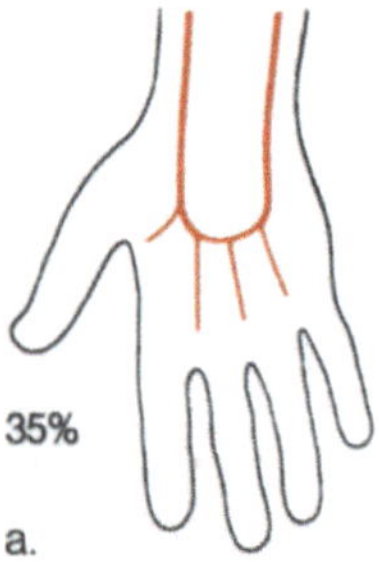

35%

a.

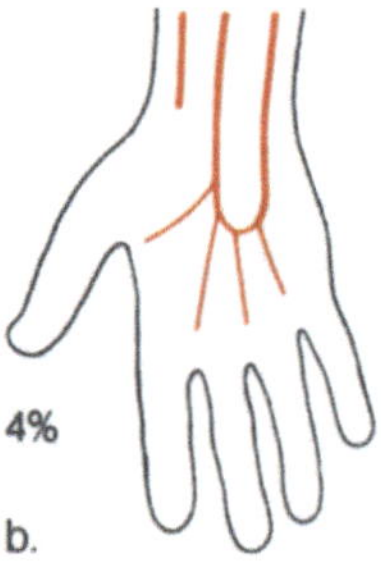

4%

b.

a. "normal" situation as described in textbooks (radioulnar type)
b. medianoulnar type

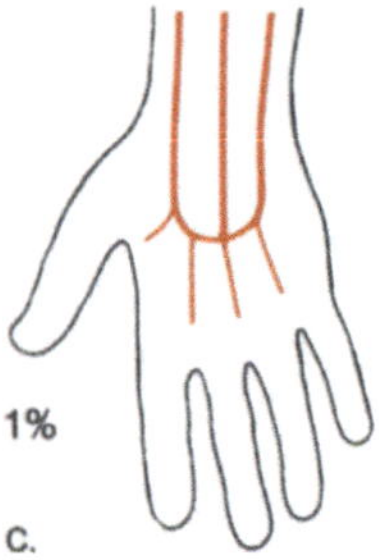

1%

c.

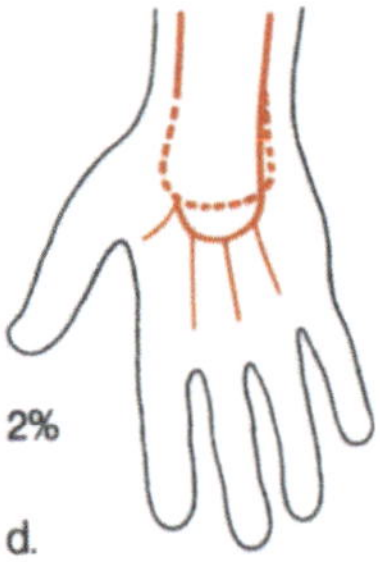

2%

d.

c. radiomedianoulnar type
d. profundoulnar type (the branch of the radial artery comes from the dorsal side or from the deep palmar arch)

2. Incomplete arch 58%

a. all four common palmar digital arteries from the ulnar artery
b. the first originates from the radial, the others from the ulnar artery

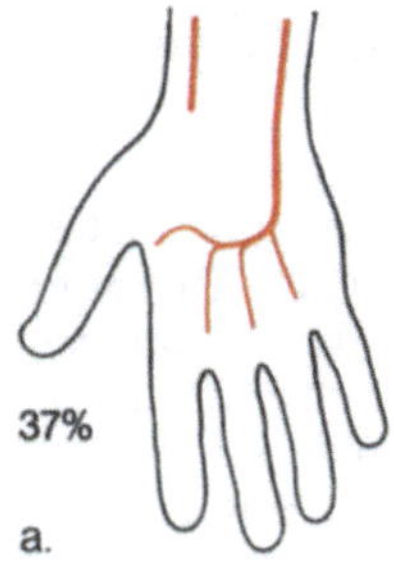

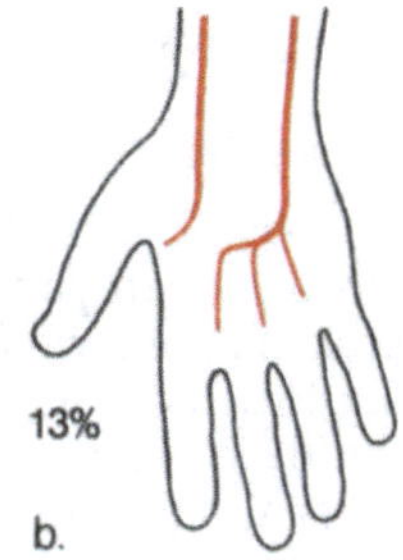

c. the first two common palmar digital arteries originate from the radial, the other two from the ulnar artery
d. the median artery continues as the second common palmar digital artery

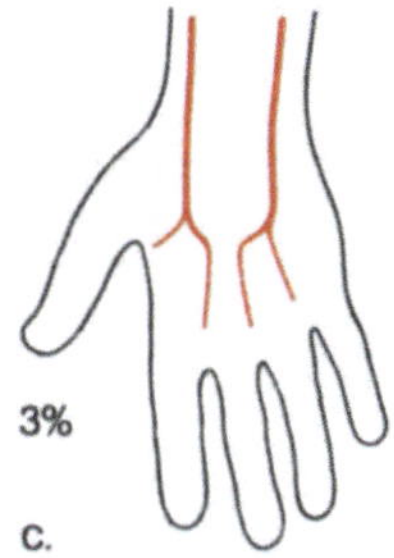

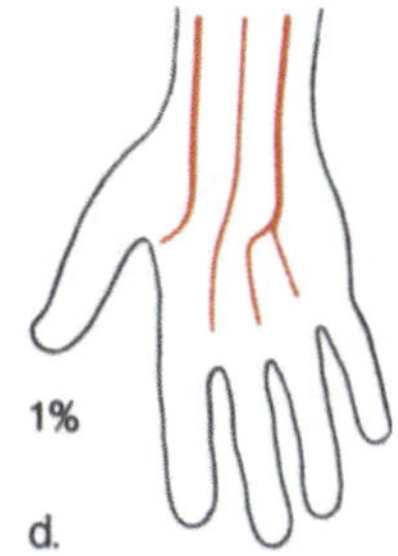

e. the first two common palmar digital arteries are branches of the median artery, the others branch from the ulnar artery

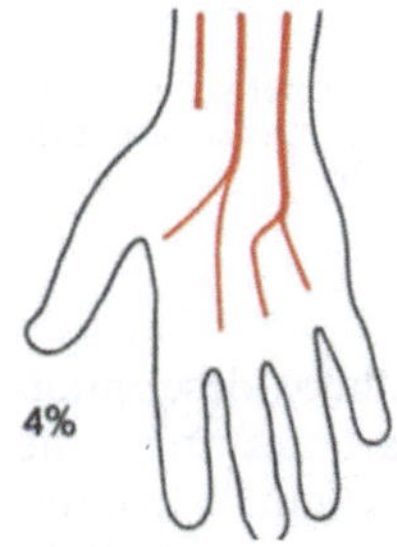

37 Deep palmar arch and palmar digital arteries

Deep palmar arch

In about 2/3 of all cases, the ulnar artery has two deep palmar branches. The more proximal branch runs to the hypothenar muscles and rarely comes into contact with the deep palmar arch (type b). The additional supply of the interosseous artery to the deep palmar branch is sometimes of practical importance: if the radial and ulnar arteries are obliterated distal to the beginning of the common interosseous artery, the hand can be sufficiently supplied with blood by the anterior interosseous artery (type c). Some case reports of this situation have been published. Since both palmar arches supply the same area, it is quite obvious that if one arch is very prominent, the other will be small [3, 15*, 103*, 125, 136, 288, 332, 350, 359, 422, 636].

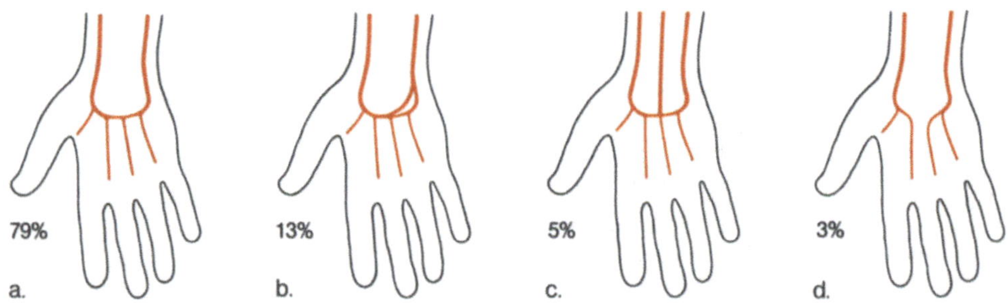

a. four common palmar digital arteries derive from the superficial palmar arch
b. three of these arteries and a proper palmar digital artery to the thumb or index finger originate from the superficial palmar arch
c. only three common palmar digital arteries
d. only two common palmar digital arteries from the superficial palmar arch

Palmar digital arteries

According to the usual textbook description the palmar metacarpal arteries from the deep arch unite with the common palmar digital arteries from the superficial palmar arch. Thus, both arches supply the fingers. However, this is valid for only 30% of all cases and in a further 10% the palmar metacarpal arteries anastomose with the proper palmar digital arteries. In the other 60%, there is no functional anastomosis [3, 11, 15*, 144, 332, 350].

Many more combinations of anomalies are possible, for individual common palmar digital arteries can be absent and some metacarpal arteries are not always present. The first palmar metacarpal artery is normally very prominent and has a proper name – the princeps pollicis artery. It also has a branch to the index finger.

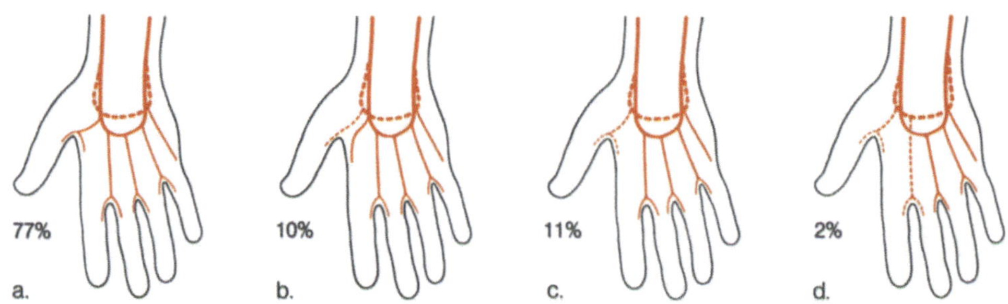

a. the ulnar artery has one deep palmar branch to the arch
b. the ulnar artery has two deep palmar branches
c. the anterior interosseous artery also supplies the deep arch
d. the deep palmar arch is incomplete

38 Arteries on the dorsal side of the hand

The net of arteries on the dorsal side of the hand is not as important as the palmar arches. The dorsal side of the fingers is supplied by the perforating branches from the palmar arteries (for the proximal parts of the fingers) and by branches of the proper palmar digital arteries for the middle and distal segments of the fingers. The arteries of the fingers follow a course similar to that of the nerves. In most cases the posterior and anterior interosseous arteries take part in the blood supply of the dorsal side of the hand. The anterior interosseous artery sends a branch through the interosseous membrane to the dorsal side, where it unites with the posterior interosseous artery [3, 15, 103*, 125, 136, 332, 636].

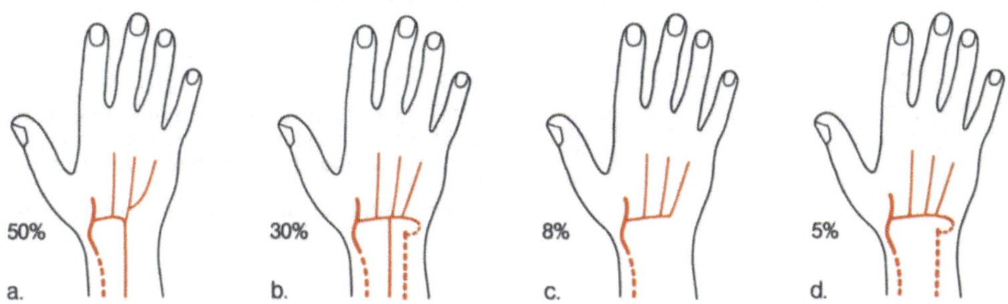

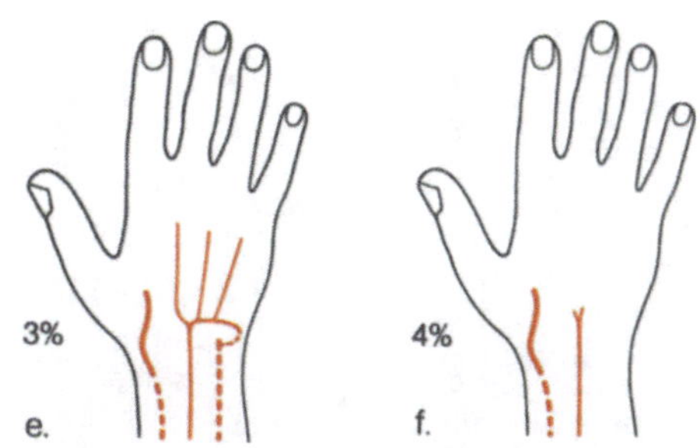

1. Dorsal arterial net of the hand

a. radiointerosseous type
b. radiointerosseoulnar type
c. radial type
d. radioulnar type
e. interosseoulnar type
f. no arterial net on the dorsal side

Radial artery

The possibilities outlined for the course of the radial artery can be further subdivided according the position of the long radial extensor muscle in relation to the tendon: in type a the radial artery almost always runs on the radial side, in type b–d often on the ulnar side.

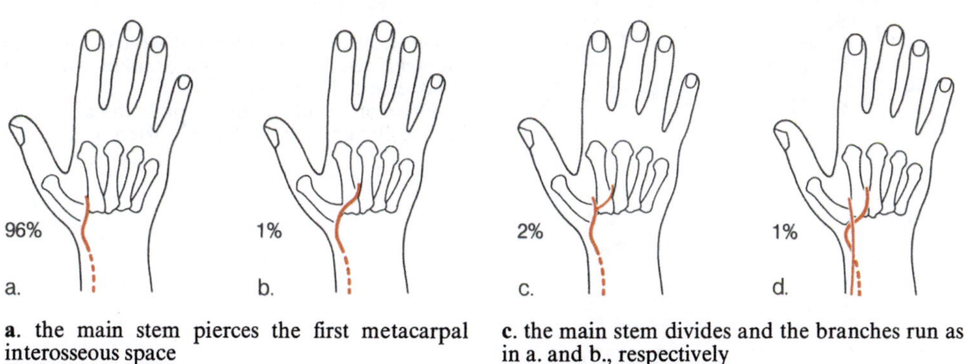

a. the main stem pierces the first metacarpal interosseous space
b. the main stem pierces the second metacarpal interosseous space
c. the main stem divides and the branches run as in a. and b., respectively
d. a superficial radial artery: the normal radial artery pierces the second metacarpal interosseous space

39 Subclavian artery

In this figure some rare combinations of trunk formations have not been included. A separate origin of the inferior thyroid, suprascapular and transverse cervical artery occurs in about 5% of all cases. A similar frequency has been described for a separate origin of the deep cervical and the supreme intercostal artery, which are the main branches of the costocervical trunk [3*, 15*, 21, 69, 113*, 164*, 284, 301, 312, 331, 412, 414, 433, 516].

1. Trunk formation of the inferior thyroid, suprascapular, transverse cervical and internal thoracic artery; the vertebral artery and the costocervical trunk are separate 85%

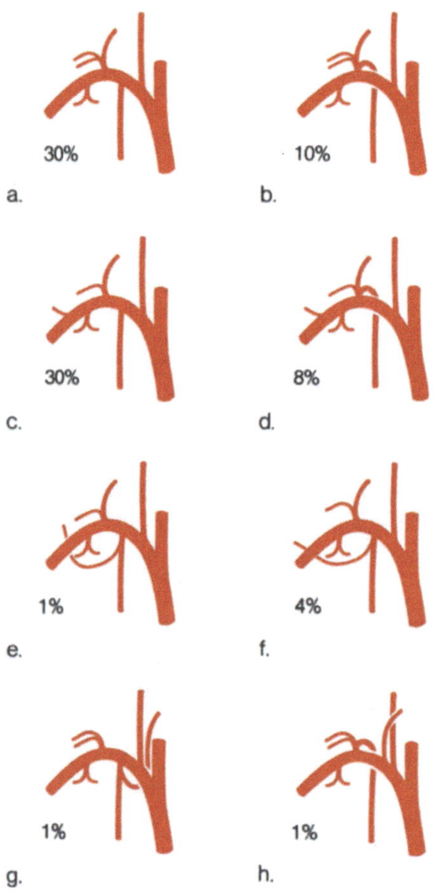

a. "normal situation" as shown in textbooks: the thyrocervical trunk is formed by the inferior thyroid, suprascapular and the transverse arteries of the neck
b. the internal thoracic artery originates from the thyrocervical trunk

c. only the inferior thyroid and the suprascapular artery form a common stem
d. common trunk of the inferior thyroid, suprascapular and internal thoracic arteries

e. two trunks: the inferior thyroid with the suprascapular artery and the transverse colli with the internal thoracic artery
f. two trunks: the inferior thyroid with the transverse colli arteries and the suprascapular with the internal thoracic arteries

g. two trunks: inferior thyroid with the internal thoracic arteries and suprascapular with the transverse colli arteries
h. common trunk of the suprascapular, transverse colli and internal thoracic arteries

2. Participation of the vertebral artery and/or the costocervical trunk 10%

a. costocervical trunk with the transverse colli artery
b. costocervical trunk with the internal thoracic artery

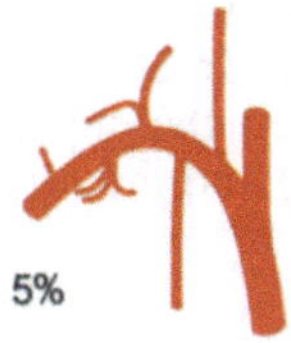

a.

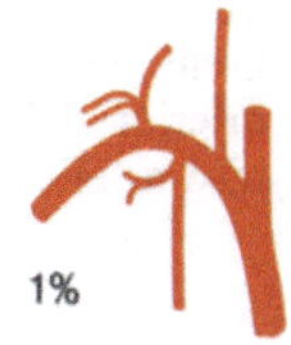

b.

c. costocervical trunk with the suprascapular artery
d. costocervical trunk with the inferior thyroid artery

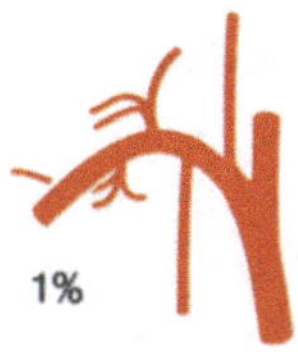

c.

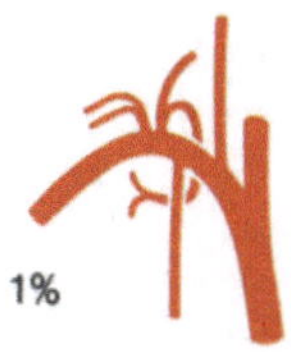

d.

e. costocervical trunk with the inferior thyroid and suprascapular arteries
f. costocervical trunk with the vertebral artery

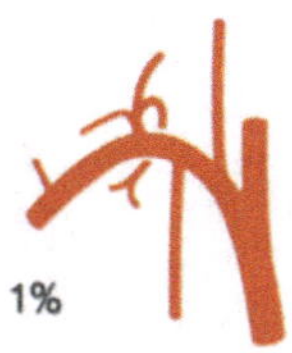

e.

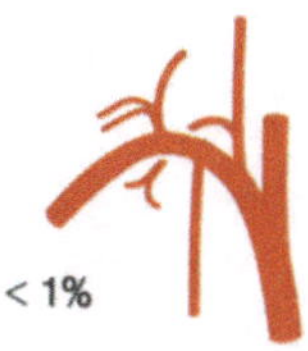

f.

g. vertebral artery with the inferior thyroid artery
h. vertebral artery with the thyrocervical trunk

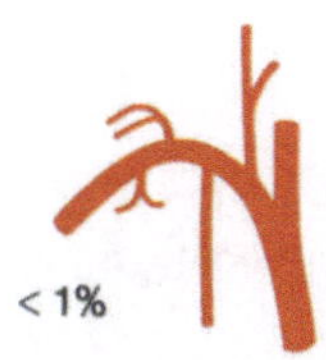

g.

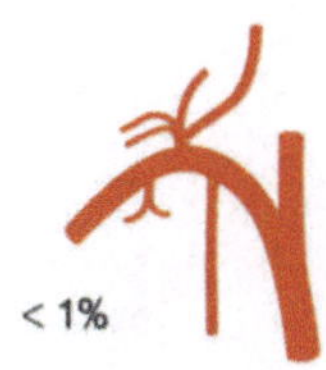

h.

40 Inferior thyroid artery

The blood supply to the thyroid gland is of great clinical relevance. In addition to the inferior thyroid arteries, the superior thyroid arteries (see p. 83) and the arteria thyroidea ima (see p. 6) show considerable variations [3*, 57, 164*, 186, 253, 312*, 331, 380, 443, 540, 579, 603, 604].

1. Origin from the subclavian artery medial to the anterior scalenus muscle [95%]

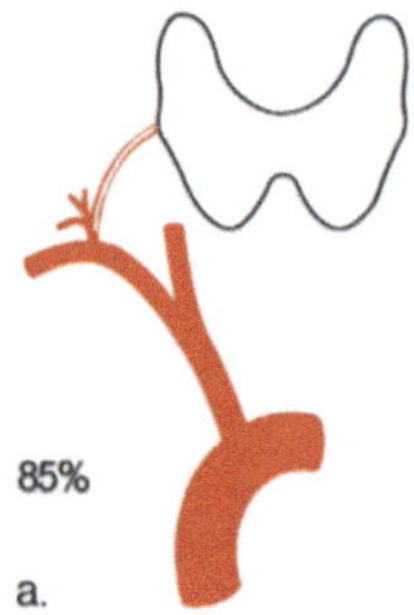

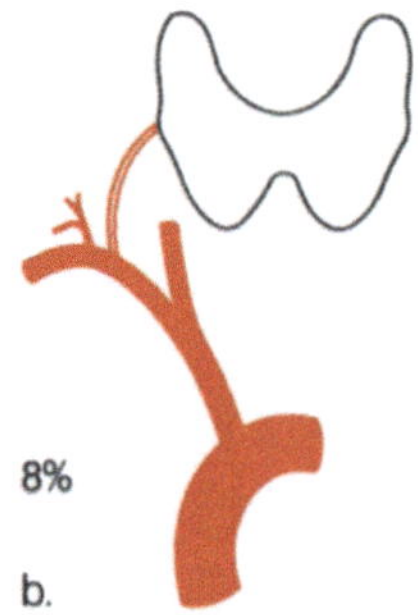

a. from the thyrocervical trunk (normal case as shown in textbooks)
b. direct from the subclavian artery

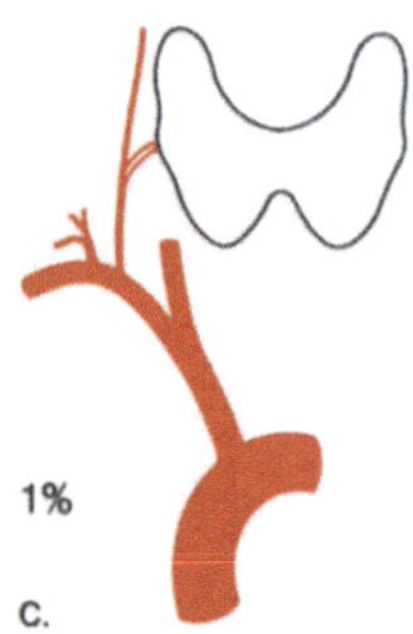

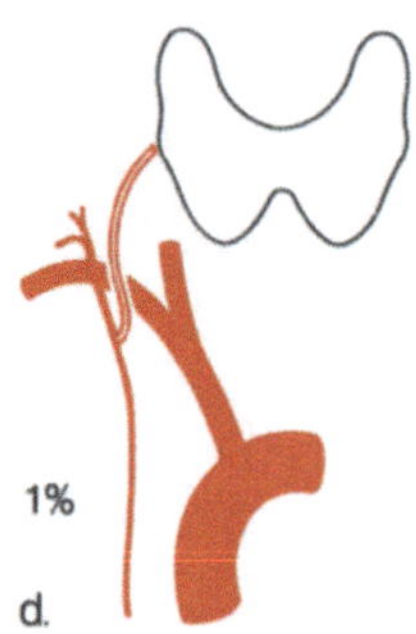

c. common stem with the vertebral artery
d. common stem with the internal thoracic artery

2. Origin from the subclavian artery, dorsal or lateral to the anterior scalenus muscle [1%]

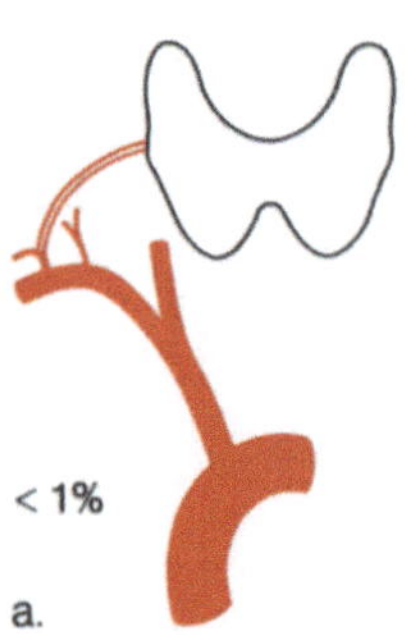

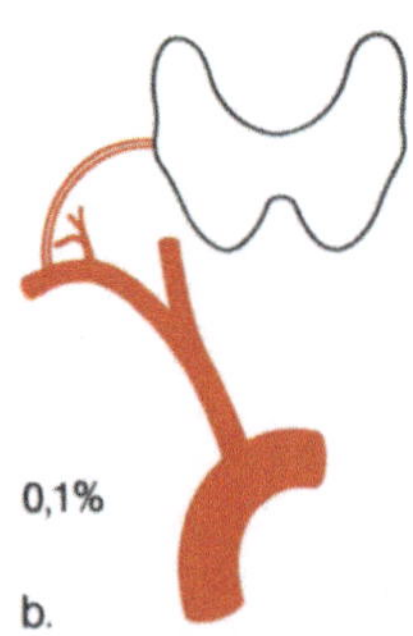

a. common origin with the suprascapular artery or the costocervical trunk
b. direct from the subclavian artery

3. Origin from the common carotid artery [1%]

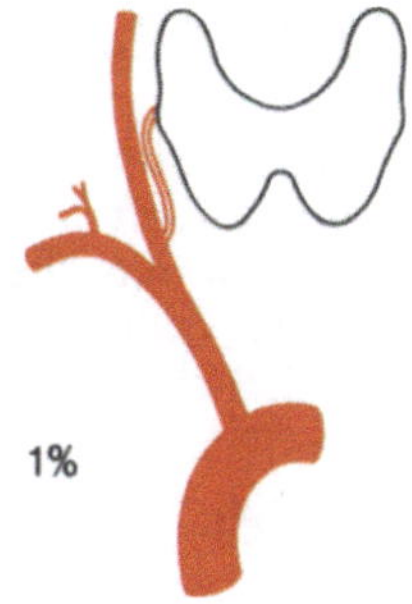

1%

in most cases on the right side, but also from the bifurcation of the brachiocephalic trunk

4. Absence of the inferior thyroid artery [3%]

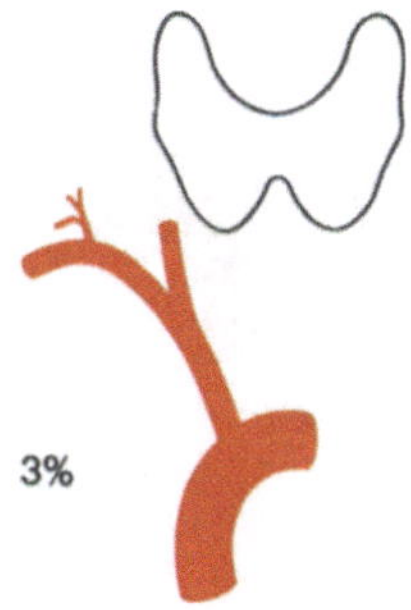

3%

the superior thyroid artery or the arteria thyroidea ima supplies the area of the inferior thyroid artery

Topographical relationship to the laryngeal nerve: [3*, 57*, 380, 603*]

a. the recurrent laryngeal nerve is ventral to the inferior thyroid artery
b. the recurrent laryngeal nerve runs between the main branches of the inferior thyroid artery

30%

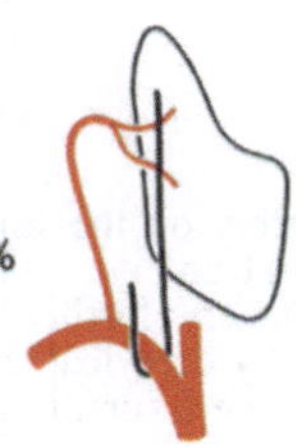

a.

45%

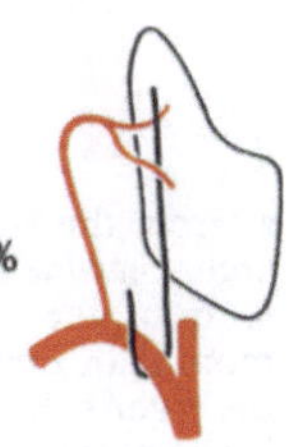

b.

c. the recurrent laryngeal nerve runs dorsally to the inferior thyroid artery
d. there is no recurrent laryngeal nerve (direct branch of the vagal nerve in cases of an arteria lusoria p. 6)

24%

c.

1%

d.

41 Vertebral artery

Although the number of anomalies on the right and left side seem to be comparable on the whole, there are major differences between sides within the different types. The left side shows a tendency to a more medial origin, the right side, to a more lateral origin [3, 22*, 93, 113, 182, 202, 266, 312*, 331, 437*, 464, 465, 536, 541, 599, 606].

1. Origin

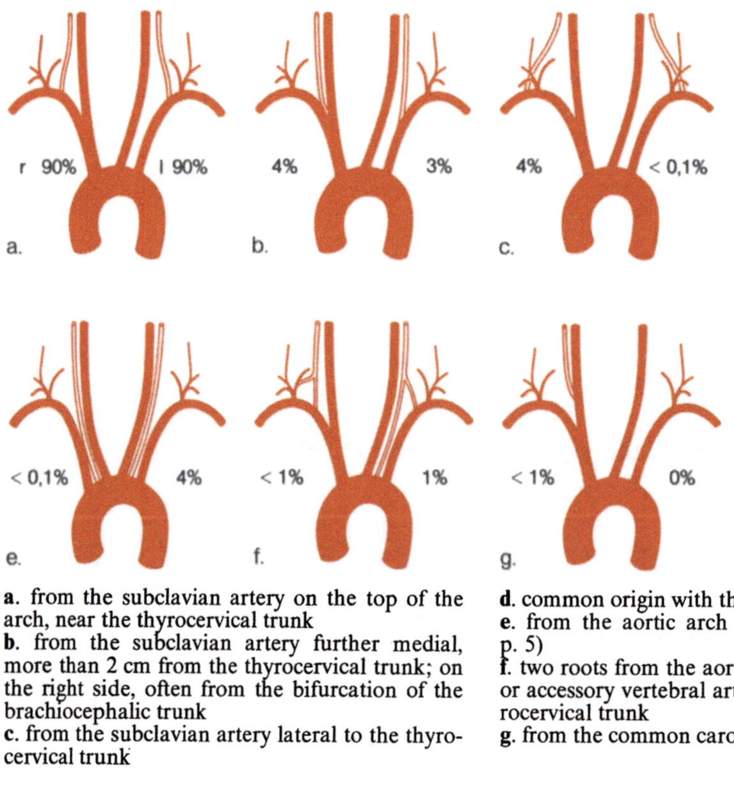

a. from the subclavian artery on the top of the arch, near the thyrocervical trunk
b. from the subclavian artery further medial, more than 2 cm from the thyrocervical trunk; on the right side, often from the bifurcation of the brachiocephalic trunk
c. from the subclavian artery lateral to the thyrocervical trunk
d. common origin with the thyrocervical trunk
e. from the aortic arch (different subtypes, see p. 5)
f. two roots from the aorta and subclavian artery or accessory vertebral arteries, also from the thyrocervical trunk
g. from the common carotid artery

2. Entrance into the cervical vertebral column

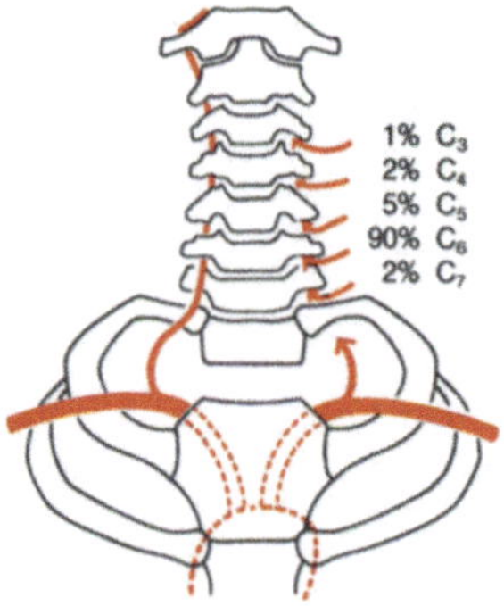

The caudalmost transverse foramen is indicated, through which the vertebral artery runs. On the left side the vertebral artery enters another transverse foramen more often than the 6th as generally stated in textbooks [3, 15*, 22*, 93].

42 External carotid artery

Bifurcation of the common carotid artery. The division of the common carotid artery has often been related to the larynx. The height of the larynx is age-dependent, resulting in different heights of division of the common carotid artery according to age. It is more reliable to relate the bifurcation to the vertebral column to avoid age dependency. Differences of up to the height of one cervical vertebra have been described between the right and left side. Arch-like divisions are more common in higher divisions and divisions at acute angles more frequent in lower divisions [3*, 312*, 333, 345*, 492*, 559].

The basis for the following figures was mainly provided by the data of Adachi [3]. The references are given for anomalies of the carotid arteries in general [3*, 78, 93, 112, 181, 312*, 315, 322a, 331, 333*, 352*, 382, 439, 529, 545, 548, 554, 639, 651] (see also aortic arch pp. 4–9) and for the branches and trunks of the external carotid artery. Many studies deal with more than one branch [3*, 260, 312*, 326*, 331, 333, 492].

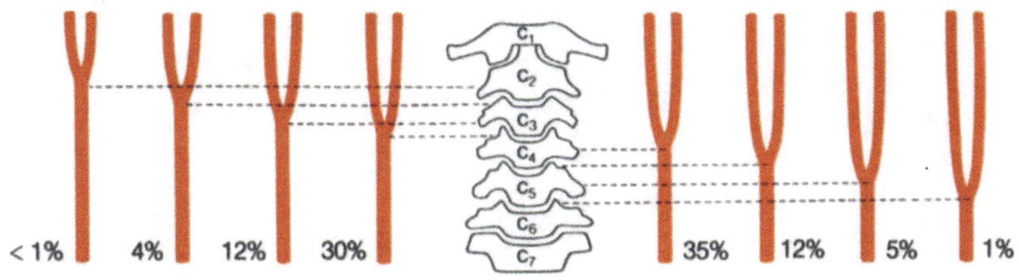

Superior thyroid, lingual and facial arteries

1. Separate origins 80%

a. all three arteries originate from the external carotid artery

b. the superior thyroid artery originates from the bifurcation of the common carotid artery

c. the superior thyroid artery originates from the common carotid artery (on the left side more often than the right, in females more often than males)

d. the superior thyroid and lingual arteries originate from the common carotid artery

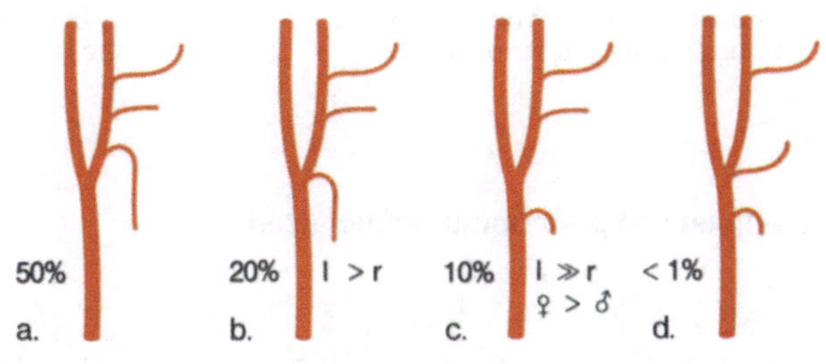

2. Trunk formation 20%

a. linguofacial trunk (always from the external carotid artery, the superior thyroid artery might derive from the common carotid artery)

b. thyrolingual trunk from the external carotid artery

c. thyrolingual trunk from the common carotid artery

d. thyrolinguofacial trunk (always from the external carotid artery)

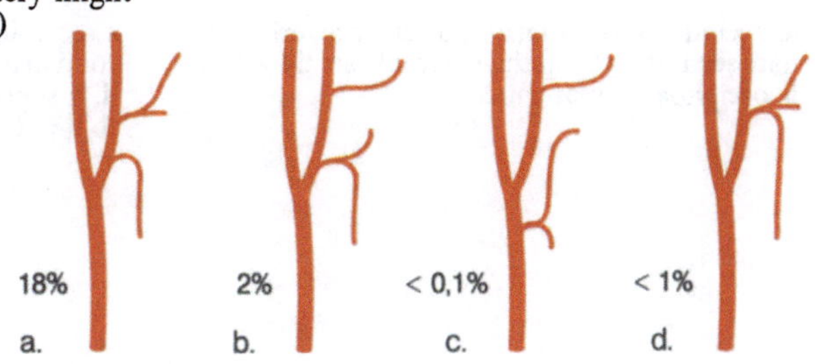

Superior laryngeal artery

1. Origin

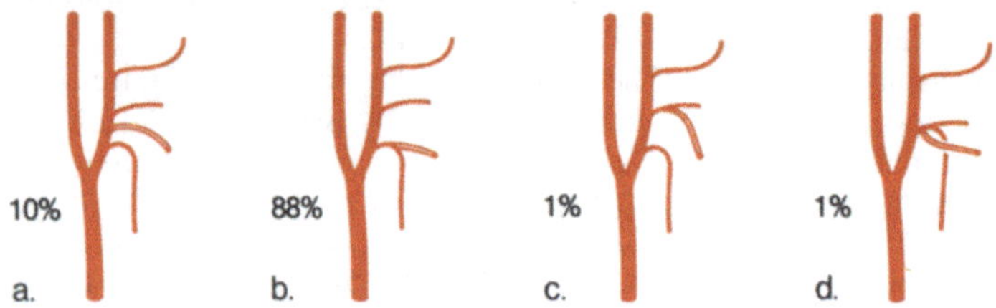

a. from the external carotid artery
b. from the superior thyroid artery
c. from the lingual artery
d. from the thyrolingual or thyrolinguofacial trunk

2. Course

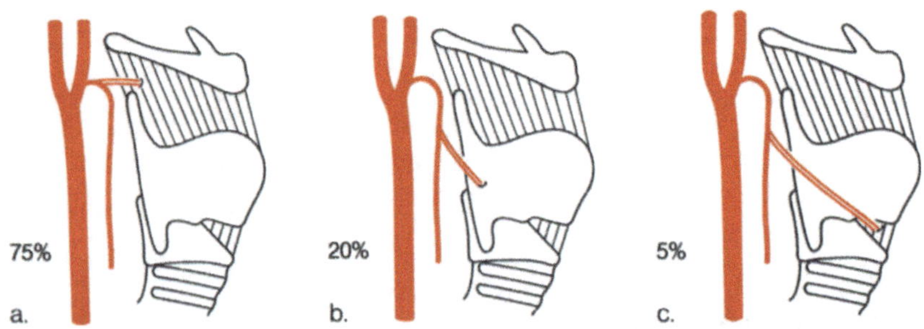

a. through the thyrohyoid membrane
b. through a thyroid foramen
c. between the thyroid and cricoid cartilage into the larynx

Occipital and posterior auricular arteries

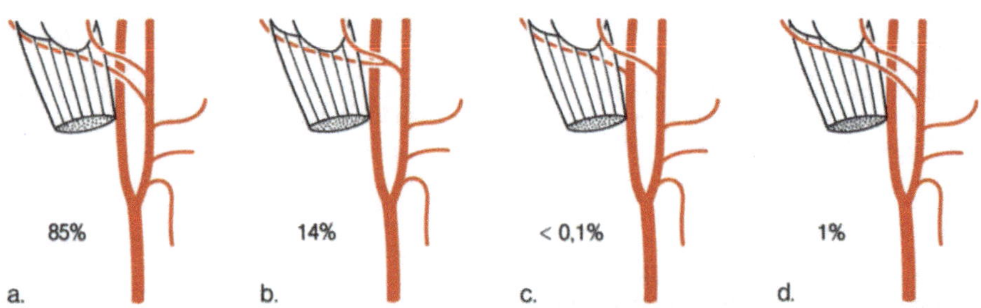

a. occipital and posterior auricular arteries originate separately from the external carotid artery
b. occipitoauricular trunk
c. occipital artery originates from the internal carotid artery
d. a superficial occipital artery runs superficially to the sternocleidomastoid muscle

Ascending pharyngeal artery

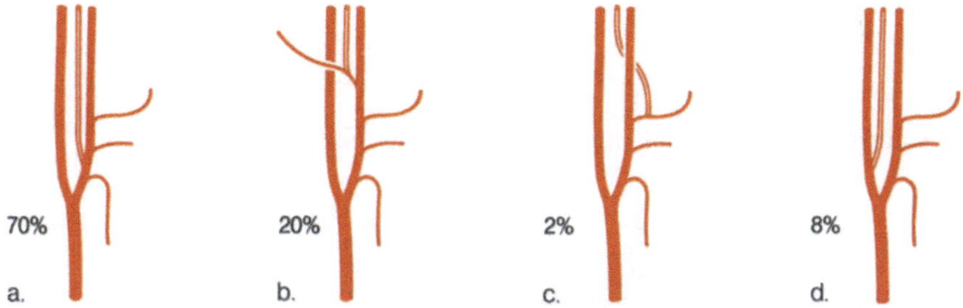

a. direct origin from the external carotid artery
b. origin from the occipital artery
c. origin from the facial artery
d. origin from the internal carotid artery

Ascending palatine artery

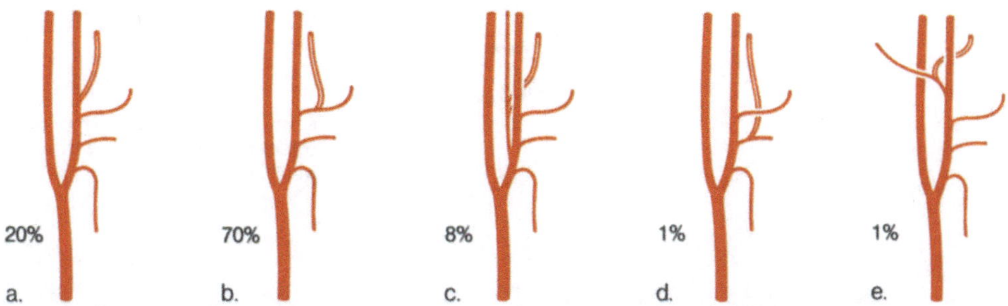

a. direct origin from the external carotid artery
b. origin from the facial artery
c. origin from the ascending pharyngeal artery
d. origin from the lingual artery
e. origin from the occipital artery

43 Maxillary artery

[3*, 9*, 44, 45, 34, 76, 295, 310, 311, 312*, 333*, 336, 368, 435, 392]

1. Course lateral to the lateral pterygoid muscle 66%

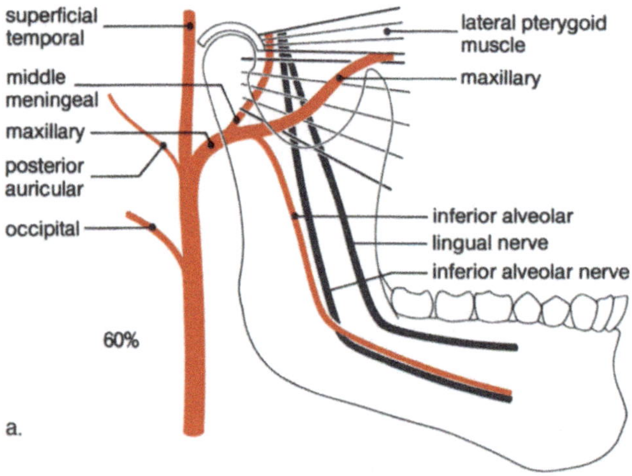

a.

a. middle meningeal artery originates proximally to the inferior alveolar artery

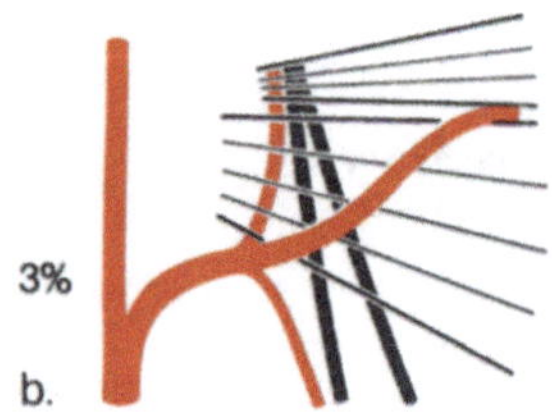

b.

b. middle meningeal artery originates opposite to the inferior alveolar artery

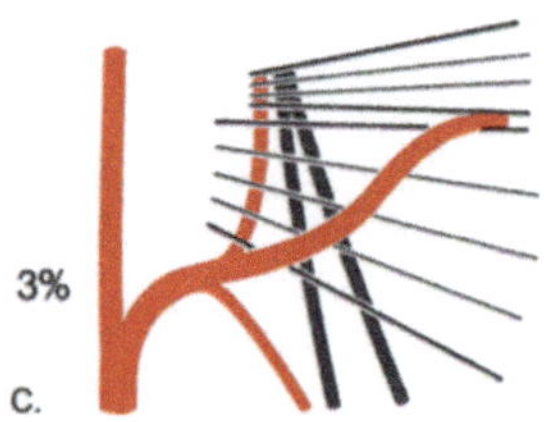

c.

c. middle meningeal artery originates distally to the inferior alveolar artery

2. Course medial to the lateral pterygoid muscle and lateral to the main branches of the mandibular nerve 21%

a. middle meningeal artery originates proximally to the inferior alveolar artery

2%

a.

b. middle meningeal artery originates opposite to the inferior alveolar artery

1%

b.

c. middle meningeal artery originates distally to the inferior alveolar artery

18%

c.

3. Course medial to the lateral pterygoid muscle and medial to or between the main branches of the mandibular nerve 13%

a. maxillary artery is medial to the lingual and inferior alveolar nerves (middle meningeal artery is always distal to the inferior alveolar artery)

6%

a.

b. maxillary artery between the lingual and inferior alveolar nerves

3%

b.

c. maxillary artery runs through a loop in the inferior alveolar nerve

4%

c.

44 Development of the arteries of the head

The arteries of the head derive from branchial arteries, the early development of which is described on page 3. The cervical segmental arteries form an anastomotic chain which later becomes the vertebral artery. The corresponding "segmental arteries" of the head (proatlantal, trigeminal, otic and hypoglossic artery) are connected with the terminal branch of the internal carotid artery, the cerebral vertebral artery, which bends in a dorsal and caudal direction.

The cervical and cerebral vertebral arteries form one continuous artery on each side and the segmental arteries disappear. Between the trigeminal and hypoglossic artery the initially paired vertebral arteries merge to form the basilar artery.

At an early stage a large area of the face is supplied by a branch of the second branchial artery, the stapedial artery. Its name derives from the stapes of the middle ear which forms around this artery. The three main branches of the stapedial artery join the three branches of the trigeminal nerve, and the supraorbital, infraorbital and mandibular arteries. Later two anastomoses are formed, one between the maxillary artery from the external carotid artery and the mandibular artery, and the other between the ophthalmic and the supraorbital arteries. The stapedial artery atrophies and the maxillary artery supplies almost the whole area of the primitive stapedial artery. Only the supraorbital artery remains connected to the ophthalmic artery.

The anomalies shown on pp. 86–87 for the maxillary artery can be explained by the embryological development. The original anastomosis between the stapedial and external carotid arteries is medial to the mandibular nerve. As a rule, a second anastomosis is formed lateral to the lateral pterygoid muscle and the first anastomosis disappears [76, 83, 244, 276, 295, 312*, 322a, 333*, 334, 335, 353*, 376, 413*, 420, 553, 589, 616, 642].

1. Persistent stapedial artery <0.1%

A persistent stapedial artery is a very rare anomaly. This artery originates from the extracranial part of the internal carotid artery, enters the skull medial to the styloid process, runs to the middle ear in an osseous canal, is surrounded by the stapes and reaches the middle cerebral fossa in front of the canal of the major petrosal nerve and ends as a middle meningeal artery [3, 221, 294, 312*, 333*, 353, 392, 427, 511, 590].

2. Anastomoses between internal carotid and basilar artery

a. Trigeminal artery <0.1%

The trigeminal artery branches from the intracranial internal carotid artery within the cavernous sinus, pierces the dorsum sellae frequently and reaches the rostral parts of the basilar artery. It is 2–3 cm long. This artery can be compared to a prominent posterior communicating artery. The increased performance of angiographies of cerebral arteries has resulted in the more frequent description of this anomaly [47, 168, 258, 268, 312, 319, 333, 353, 417, 533, 534, 592, 607, 666].

b. Hypoglossic artery ≪0.1

The persistent hypoglossic artery has been described less frequently than the trigeminal artery. This vessel originates from the extracranial part of the internal carotid artery, enters the skull via the hypoglossic canal and ends in the caudal part of the basilar artery instead of a vertebral artery [12, 13, 62, 77, 84, 169, 198, 239, 353, 508, 551, 592, 660].

c. Proatlantal artery ≪0.1%

This artery is comparable to the hypoglossic artery, but it runs between the atlas and occipital bone and enters the skull via the great foramen. This vessel has also been named the primitive suboccipital artery [13, 55, 312, 333, 353, 451, 466, 487, 503, 623].

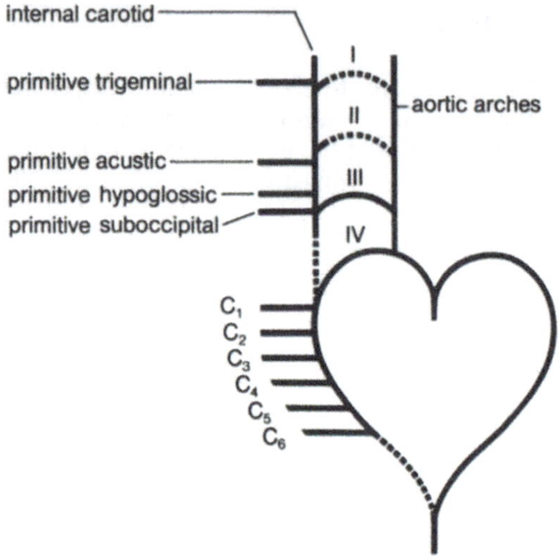
internal carotid
primitive trigeminal
primitive acustic
primitive hypoglossic
primitive suboccipital
I
II
III
IV
aortic arches
C1
C2
C3
C4
C5
C6

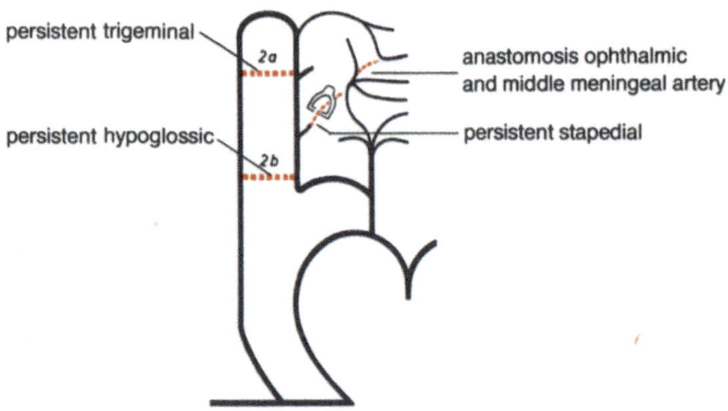
persistent trigeminal
2a
persistent hypoglossic
2b
anastomosis ophthalmic
and middle meningeal artery
persistent stapedial

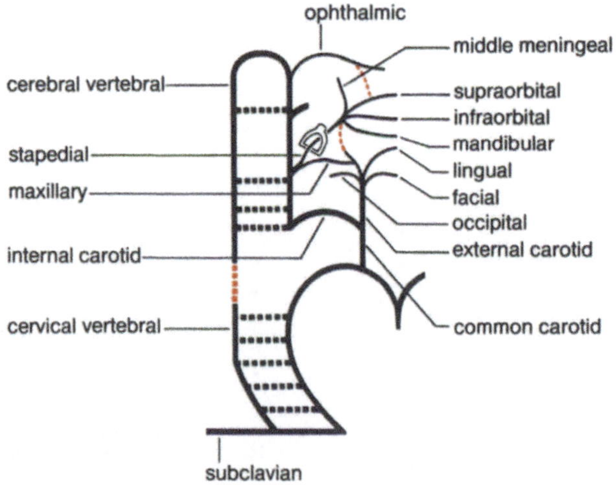
ophthalmic
middle meningeal
cerebral vertebral
supraorbital
infraorbital
mandibular
stapedial
lingual
maxillary
facial
occipital
internal carotid
external carotid
cervical vertebral
common carotid
subclavian

45 Ophthalmic artery

The practical importance of an anastomosis between the ophthalmic and middle meningeal artery is that branches of the external and internal carotid artery anastomose. The frequency of the origin of the middle meningeal artery from the ophthalmic artery seems to reflect racial differences, for instance in skulls of people from Papua the foramen spinosum is absent in about 10% [3, 44*, 76, 295, 312*, 332 a, 332 b*, 333].

1. Connections between the ophthalmic and middle meningeal arteries

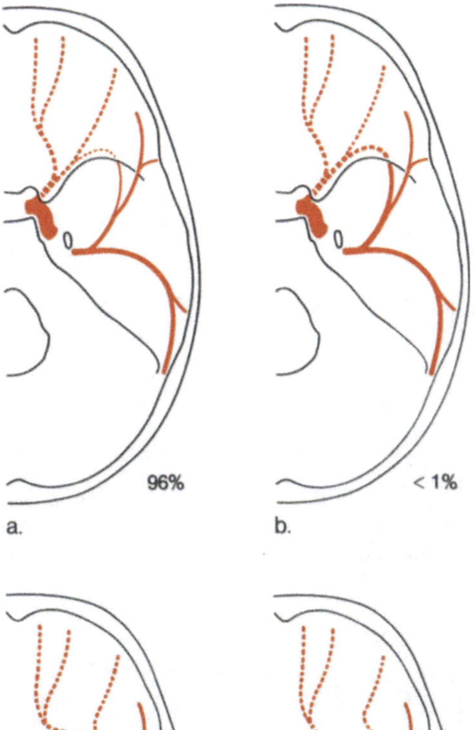

a. b.

a. "normal" situation: only a minor anastomosis between the ophthalmic and middle meningeal arteries
b. large anastomosis: the ophthalmic artery originates with comparably strong roots from the internal carotid and middle meningeal arteries

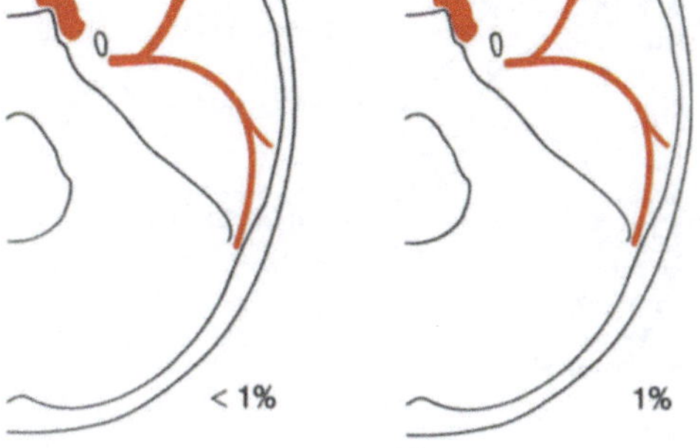

c. d.

c. the ophthalmic artery is a branch of the middle meningeal artery
d. the lacrimal artery is a branch of the middle meningeal artery

e. the frontal branch of the middle meningeal artery originates from the ophthalmic artery
f. the middle meningeal artery derives from the ophthalmic artery; the foramen spinosum is absent

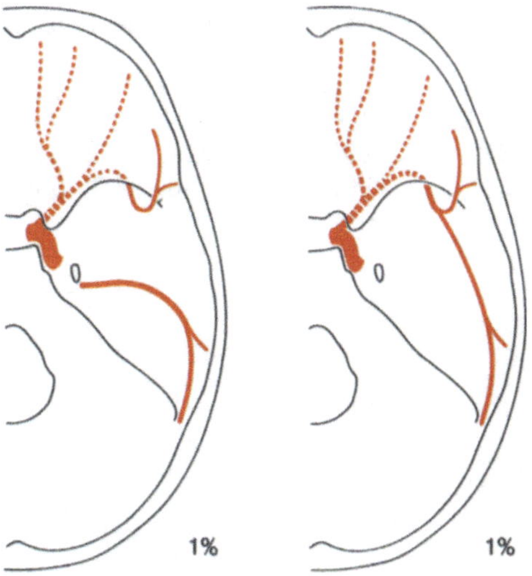

e. f.

2. Position of the ophthalmic artery relative to the optic nerve

a. the ophthalmic artery crosses over the optic nerve
b. the ophthalmic artery crosses under the optic nerve

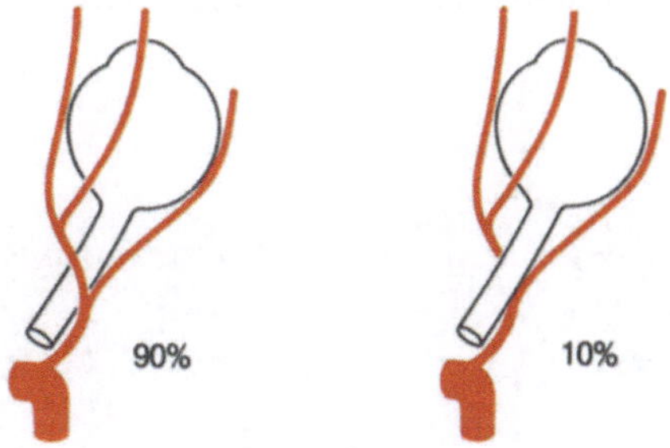

a. b.

46 Cerebral arterial circle (circle of Willis)

There are many anomalies of the arteries at the base of the skull [2, 79, 85, 207, 261*, 297*, 303, 309*, 312*, 322a, 322b, 328*, 333*, 353, 353a, 411, 462, 589, 597]. Some studies deal with anomalies of the internal carotid artery [181, 210, 299, 322a, 322b, 490, 524, 609, 625] and others with the circle of Willis [3, 101*, 118, 182, 312*,

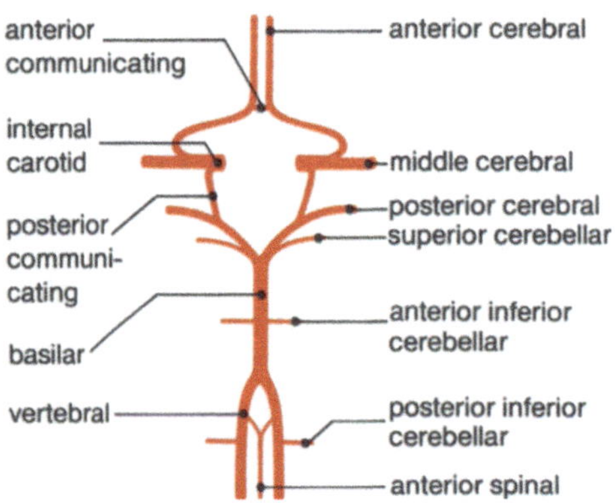

1. Anterior part of the cerebral circle

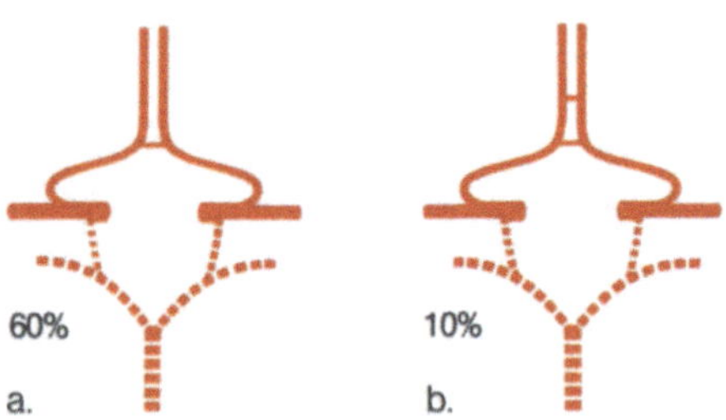

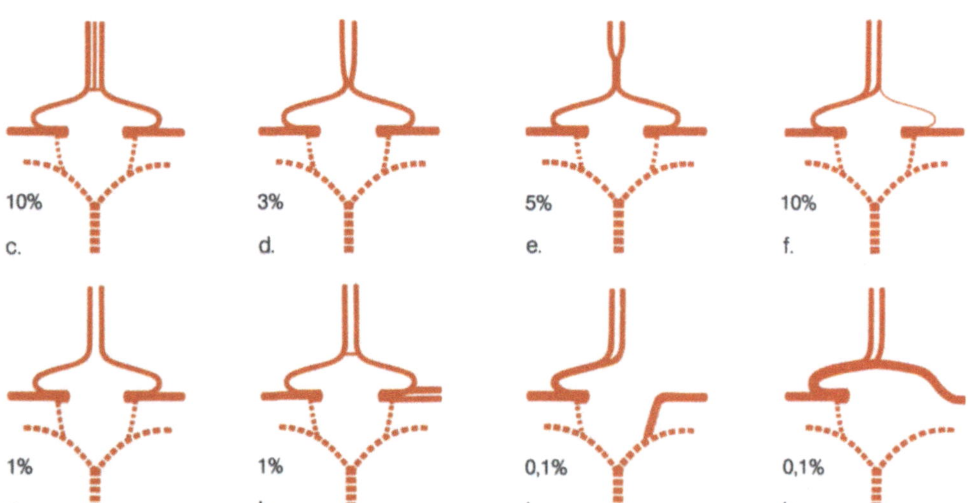

a. "normal" situation
b. two anterior communicating arteries
c. the median artery of the corpus callosum originates from the anterior communicating artery
d. instead of the anterior communicating artery, a lateral anastomosis between the two anterior cerebral arteries is present
e. both anterior cerebral arteries combine partly in a common trunk
f. both anterior cerebral arteries originate from one internal carotid artery
g. absence of the anterior communicating artery
h. accessory middle cerebral artery
i. absence of one internal carotid artery; the middle cerebral artery originates from the basilar artery
k. absence of one internal carotid artery; the middle cerebral artery originates from the other internal carotid artery

322 a, 322 b, 341*, 388, 407, 459*, 498, 514*, 542, 558] or its branches as the middle cerebral artery [240, 278, 322 a, 322 b, 324, 608, 652] or cerebellar arteries [327*, 429*]. For didactic reasons, the anomalies of the anterior area [109, 139, 216, 322 a, 322 b, 325*, 522], posterior area [322 a, 322 b, 556] and the basilar artery are depicted separately. These anomalies can be combined in different individuals and result in many different types. The frequencies given here are a compromise between radiological and anatomical studies. Vessels were termed "small" or "absent" (e.g. type 2 b), when only negligible collateral circulation took place via such a route.

2. Posterior part of the cerebral circle

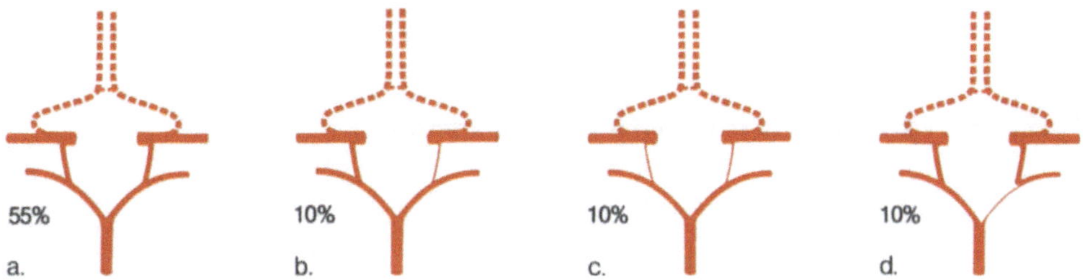

a. "normal" situation
b. the posterior communicating artery is absent or very small on one side
c. the posterior communicating artery is absent or very small on both sides
d. on one side the posterior cerebral artery originates from the internal carotid artery
e. on both sides the posterior cerebral arteries originate from the internal carotid artery
f. on one side the posterior communicating artery is absent, on the other side the posterior cerebral artery comes from the internal carotid artery

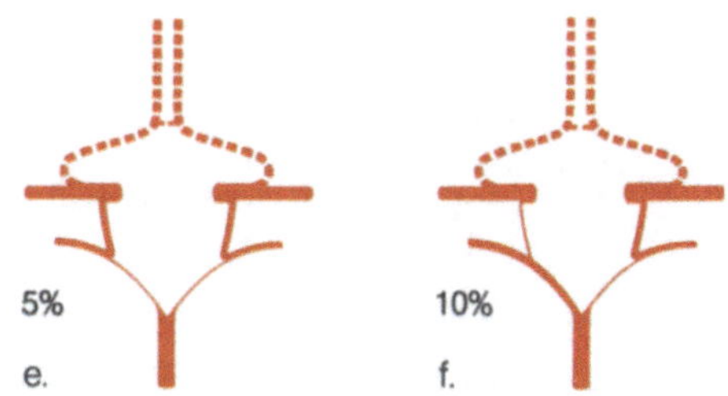

3. Basilar artery

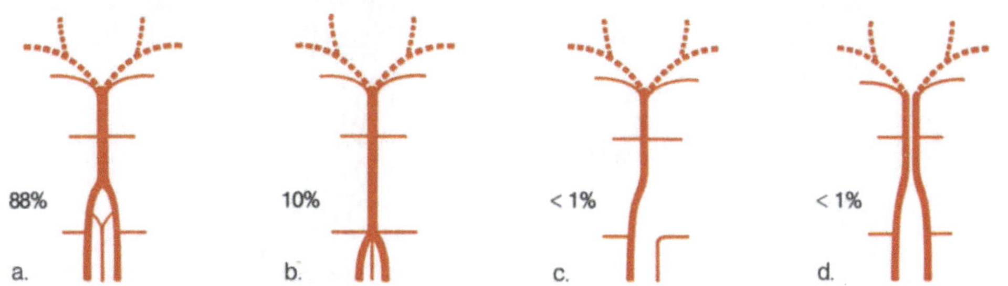

a. "normal" situation
b. both vertebral arteries unite in an extremely caudal position
c. the basilar artery is the continuation of only one vertebral artery, the other vertebral artery ends as the posterior inferior cerebellar artery in most cases.
d. two basilar arteries, but otherwise normal branches
e. "island formation" of the basilar artery
f. an anastomosing network instead of a basilar artery

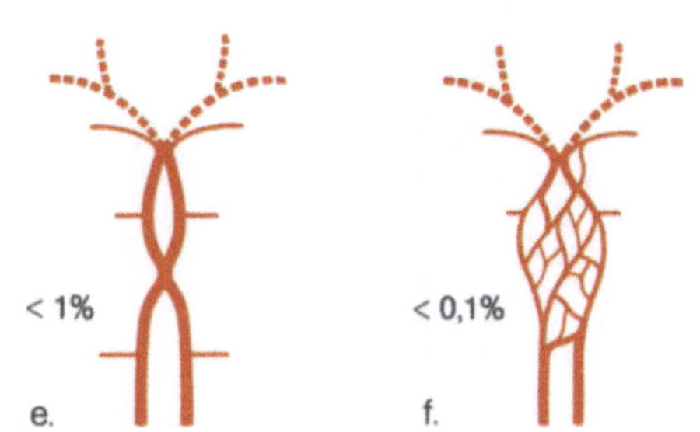

47 Arteries of the spinal cord

The spinal cord is supplied by three longitudinal arterial trunks, i.e. the anterior spinal artery and the two posterolateral trunks. The anterior spinal artery is formed by two roots from the vertebral arteries, and anterior radicular arteries reinforce the longitudinal channel at various levels; in the cervical part the anterior radicular arteries arise from the vertebral artery, thyrocervical and costocervical trunks. Branches from intercostal and lumbar arteries join the longitudinal system at the thoracic and lumbar segments. Usually one artery is larger than the other, i.e., the great anterior medullary artery (artery of Adamkiewicz radicularis magua), which is more

1 a and 1 b. Typical origin of the anterior radicular arteries

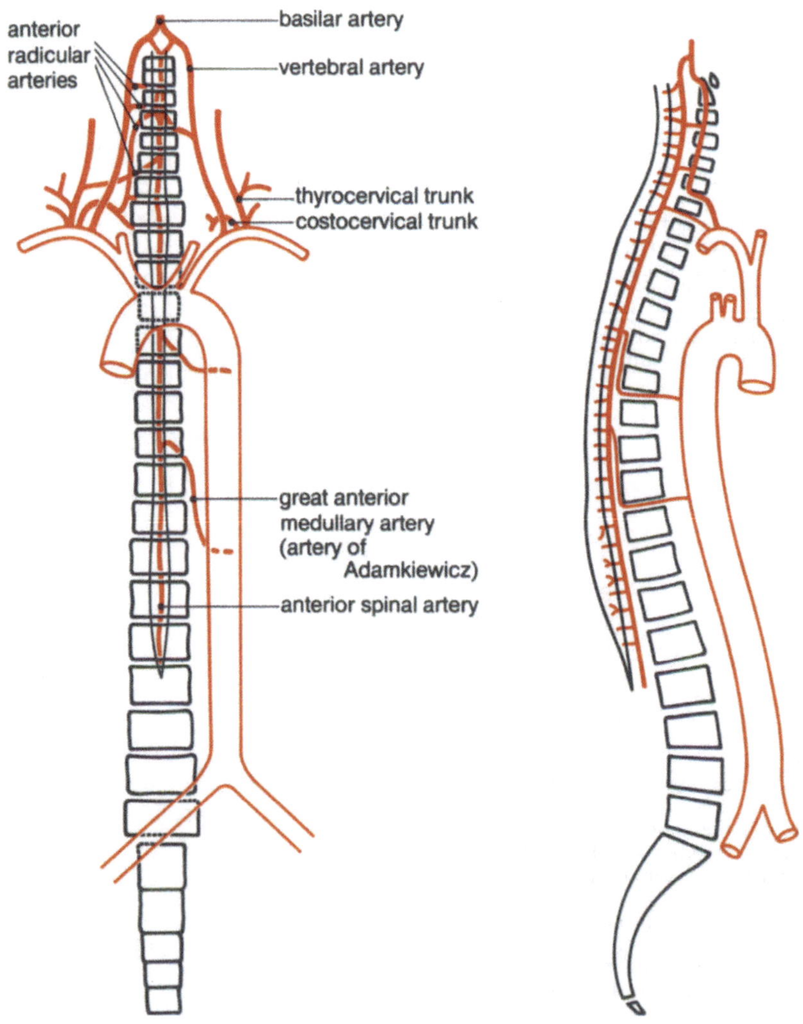

often found on the left than on the right side and usually between the levels Th9 and L2. Variations of the dorsal radicular arteries are not shown in the figure. The great variability in the number of radicular arteries, the prevalence of the left side in the thoracic and lumbar parts and the level of the radicular arteries has been explained by the embryological segmental blood supply and the preferential left side, by the more direct blood flow from the aorta to the left radicular arteries [547]. Paraplegia following orthopedic surgery of the spine or operation of the aorta stresses the clinical relevance of the anomalies of the spinal cord blood supply [75, 92, 100, 129*, 130*, 133*, 134, 135*, 166, 178, 281a*, 323*, 342*, 343, 369, 370, 444*, 489*, 537, 547*, 626, 643*].

2. Variable origin of the great anterior medullary artery (artery of Adamkiewicz)

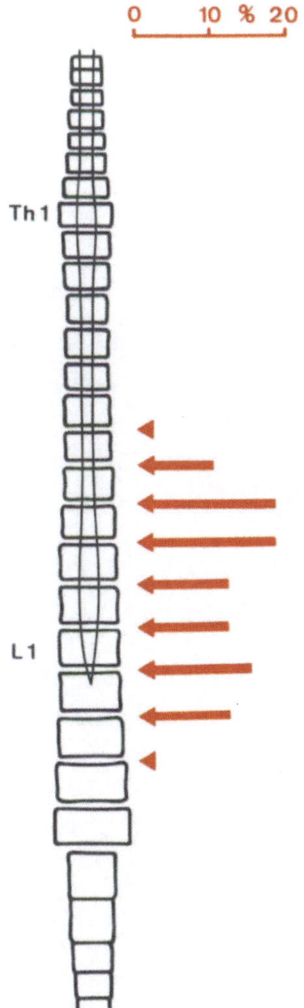

Literature

1. Abadia-Fenoll F. Über eine äußerst stark geschlängelte und verlängerte Arteria lienalis. Anat Anz 1964;115:339–44.
2. Abrams HL, ed. Abrams angiography, vascular and interventional radiology. 3rd ed. Boston: Little Brown Co, 1983, Vol. I, II, III.
3. Adachi B. Das Arteriensystem der Japaner. Bd. 2. Kyoto: Maruzen, 1928.
4. Ahmed SH, El-Rakhawy MT, Abdalla A, Harrison RG. A new conception of coronary artery preponderance. Acta Anat (Basel) 1972;83:87–94.
5. Albesi EJ. Abnormaler Ursprung einer Bronchialarterie. Anat Anz 1966;119:104–7.
6. Alexander RW, Griffith GC. Anomalies of the coronary arteries and their clinical significance. Circulation 1956;14:800–5.
7. Allan FD. An analysis of the clavicothoracic visceral branches of the vague and the sympathetic trunk in the presence of an anomalous right subclavian artery. Anat Rec 1958;132:71–80.
8. Allard JR, Williams RL, Dobell AR. Interrupted aortic arch: Factors influencing prognosis. Ann Thorac Surg 1976;21:243–6.
9. Allen WE, Kier EL, Rothman SLG. The maxillary artery: normal arteriographic anatomy. Am J Roentgenol 1973;118:517–27.
10. Allwork SP. A spectrum of normal coronary artery distribution in man. Anat Clin 1979;1:311–9.
11. Al-Turk M, Metcalf WK. A study of the superficial palmar arteries using the Doppler ultrasonic flowmeter. J Anat 1984; 138:27–32.
12. Anderson M. Persistent primitive hypoglossal artery with basilar aneurysm. J Neurology 1976;213:377–81.
13. Anderson RA, Sondheimer FK. Rare carotid-vertebrobasilar anastomoses with notes on the differentiation between proatlantal and hypoglossal arteries. Neuroradiology 1976;11:113–8.
14. Andrianov AV, Chernyshev VN, Blinnichev NM. Surgical anatomy of the left colic artery. Khirurgiia (Mosk) 1983;83:66–70.
15. Anson BJ. An atlas of human anatomy. 2nd ed. Philadelphia: Saunders, 1963.
16. Anson BJ, Cauldwell EW, Pick JW, Beaton LE. The blood supply of the kidney, suprarenal gland and associated structures. Surg Gynec Obstet 1947;84:313–20.
17. Anson BJ, Kurth LE. Common variations in the renal blood supply. Surg Gynec Obstet 1955;100:156–62.
18. Anson BJ, McVay CB. The topographical positions and the mutual relations of the visceral branches of the abdominal aorta. A study of 100 consecutive cadavers. Anat Rec 1937;67:7–15.
19. Antoine M, Pierson M, Kersauson MC. Les images radiologiques de l'oesophage infantile dans les anomalies des arcs aortiques. J Radiol Électrol Méd Nucléaire 1958; 39:174–80.
20. Arciniegas E, Hakimi M, Hertzler JH, Farooki ZQ, Green EW. Surgical management of congenital vascular rings. J Thorac Cardiovasc Surg 1979;77:721–7.
21. Ardito JM, Tucker GF, Ossoff RH, DeLeon SY. Innominate artery compression of the trachea in infants with reflex opnea. Ann Otol Rhinol Laryngol 1980;89:401–5.
22. Argenson C, Francke JP, Sylla S, Dintimille H, Papasian S, di Marino V. The vertebral arteries (segments V1 and V2). Anat Clin 1980;2:29–41.
23. Armstrong BG, Hunt TH, Price CW, Resnick MI. Common origin of inferior mesenteric and accessory renal artery. Urology 1979;14:298–9.
24. Ashley FL, Anson BJ. The hypogastric artery in American whites and negroes. Am J Phys Anthropol 1941;28:381–94.
25. Askenazi J, Nadas AS. Anomalous left coronary artery originating from the pulmonary artery: report on 15 cases. Circulation 1975;51:976–87.
26. Ayoub SF. Arterial supply to the human rectum. Acta Anat 1978;100:317–27.
27. Aytaç A, Özme S, Sarikayalar F, Saylam A. Pulmonary artery sling. Ann Thorac Surg 1976;22:596–9.
28. Bachmann D, Haasner E. Truncus intercosto-renalis. Fortschr Geb Röntgenstr Nuklearmed 1965;102:712–3.
29. Bachmann D, Haasner E. Die Nierenarteriographie unter besonderer Berücksichtigung der Nierenarterienanomalien. Fortschr Geb Röntgenstr Nuklearmed 1965; 103:71–6.
30. Bachmann D, Schaefer P. Nierenmißbildungen: Angiographische Befunde. Fortschr Geb Roentgenstr Nuklearmed. 1967; 107:50–9.
31. Bahnson HT. Symposium on pediatric surgery; coarctation of aorta and anomalies of aortic arch. Surg Clin N Am 1952; 32:1313–26.

32. Barbin JY, Visset J, Kerninon R, Fauvy A, Leborgne J, Pannier M. Variations anatomiques de l'artère colique supérieure droite. Arch Anat Path (Paris) 1972;20:373–6.
33. Barry A. The aortic arch derivatives in the human adult. Anat Rec 1951;111:221–38.
34. Baumel JJ, Beard DY. The accessory meningeal artery of man. J Anat 1961; 95:386–402.
35. Baumstark AE, Gordon DH, Haller JO, Martin EC. The left aortic arch and right descending aorta. Br J Radiol 1977; 50:788–92.
36. Bayindir S. Zur selektiven Angiographie der A. coeliaca und der Mesenterica superior. Fortschr Geb Röntgenstr Nuklearmed 1967;107:189–99.
37. Beer PM de. The profunda femoris and circumflex femoral arteries in the South African Bantu-speaking Negro. S Afr J Med Sci 1965;30:1–10.
38. Bein G, Nötges-Borgwardt J. Aortenbogenanomalien aus entwicklungsgeschichtlicher Sicht. Monatsschr Kinderheilkd 1978; 126:362–5.
39. Bein G, Wolf O. Zum Krankheitsbild des Arcus aortae duplex im Säuglings- und Kindesalter. Z Kardiol 1975;64:547–61.
40. Benson EA, Page RE. A practical reappraisal of the anatomy of the extrahepatic bile ducts and arteries. Br J Surg 1976; 63:853–60.
41. Benson PA. Anomalous aortic origin of the coronary artery with sudden death: case report and review. Am Heart J 1970;79:254–7.
42. Benton RS, Cotter WB. A hitherto undocumented variation of the inferior mesenteric artery in man. Anat Rec 1963;145:171–3.
43. Bermudez GA, Abdelnur R, Midell AI, Replogle R. Anomalous origin of the right coronary artery from the pulmonary artery with large left-to-right shunt (anomalous right coronary artery). Cathet Cardiovasc Diagn 1979;5:371–84.
44. Berthelot JL, Hureau J. Clinical anatomical study of the macroscopic anastomoses of the ophthalmic artery in the periorbital region. Anat Clin 1982;3:271–8.
45. Biermann H. Die chirurgische Bedeutung der Lagevariationen der Arteria maxillaris. Anat Anz 1943;94:289–309.
46. Billig DM. Congenital anomalies of the thoracic aorta and its branches: Pathophysiology, clinical course and management. Prog Cardiovasc Dis 1973;16:43–67.
47. Binet EF, Young RF. Bilateral persistent trigeminal arteries. Case report. J Neurosurg 1977;47:619–22.
48. Binet JP, Langlois J. Aortic arch anomalies in children and infants. J Thorac Cardiovasc Surg 1977;73:248–52.
49. Björk L. Anastomoses between the coronary and bronchial arteries. Acta Radiol Diagn 1966;4:93–6.
50. Björk L. Angiographic demonstration of extracardial anastomoses to the coronary arteries. Radiology 1966;87:274–7.
51. Blagotic M, Ilic A, Radonjic V. The final bifurcation of the left colic artery. Folia Morphol 1980;28:275–7.
52. Blair CB, Nandy K. Persistence of the axis artery of the lower limb. Anat Rec 1965;152:151–72.
53. Blake HA, Manion WC, Maitingly TW, Baroldi G. Coronary artery anomalies. Circulation 1964;30:927–40.
54. Blieden LC, Schneeweiss A, Deutsch V, Neufeld HN. Right aortic arch with left descending aorta (circumflex aorta). Roentgenographic diagnosis. Pediatr Radiol 1978;6:208–10.
55. Bloch S, Danziger J. Proatlantal intersegmental artery. Neuroradiology 1974;7:5–8.
56. Bloda E, Sierocinski W, Kling A. Variation of the Arteria profunda femoris in man. Folia Morph (Warszawa) 1982;41:123–31.
57. Blondeau P. Rapports chirurgicaux du nerf récurrent et de l'artère thyroïdienne inférieure (d'après 250 dissections chirurgicales). J Chir (Paris) 1971;102:397–414.
58. Blumenthal S, Ravitch MM. Seminar on aortic vascular rings and other anomalies of the aortic arch. Pediatrics 1957;20:896–906.
59. Bönner G, Dreesbach H, Helber A, Kaufmann W. Hypertonus und multiple Nierenarterien. Dtsch Med Wochenschr 1978; 103:345–9.
60. Bogomolova AP. Arterial arches in the area of the pancreas head in man. Arch Anat (Mosk.) 1958;35:60–3.
61. Bogren HG, Porter BA. Three aortic arch anomalies: a review of the literature and proposal of a new classification. Cardiovasc Intervent Radiol 1980;3:19–23.
62. Bohmfalk GL, Story JL. Aneurysms of the persistent hypoglossal artery. Neurosurgery 1977;1:291–6.
63. Boijsen E. Angiographic studies of the anatomy of single and multiple renal arteries. Acta Radiol Diagn (Stockh) 1959, Suppl. 183:1–35.
64. Boijsen E. Selective pancreatic angiography. Br J Radiol 1966;39:481–7.
65. Boijsen E, Olin T. Zöliakographie und Angiographie der A. mesenterica superior. Ergeb Med Strahlenforsch Neue Folge 1964;1:112–42.

66. Bookstein JJ. Aberrant left coronary artery. Am J Röntgenol 1964;91:515–28.
67. Bord RA de, Vonachen JR, Cohen MH. Double aortic arch. Surgery 1955; 38:585–92.
68. Borell U, Fernström J. The ovarian artery. An arteriographic study in human subjects. Acta Radiol Diagn (Stockh) 1954; 42:253–65.
69. Bosniak MA. An analysis of some anatomic roentgenological aspects of brachiocephalic vessels. Am J Roentgenol 1964;91:1222–31.
70. Botenga ASJ. Selective bronchial and intercostal arteriography. Leiden: Stenfert Kroese, 1970.
71. Bourdon JL, Hoeffel JC, Worms AM, Picard L, Pernot C. The cervical aortic arch: a case with diffuse arterial dysplasia and neurocutaneous angiomatosis. Pediatr Radiol 1981;10:143–50.
72. Bower EB, Smullens SN, Parke WW. Clinical aspects of persistent sciatic artery: Report of two cases and reveiw of the literature. Surgery 1977;81:588–95.
73. Boyle WF, Shaw CC. Right-sided aortic arch. New Engl J Med 1957;256:392–5.
74. Braunlin E, Peoples WM, Freedom RM, Fyler DC, Goldblatt A, Edwards JE. Interruption of the aortic arch with aortico-pulmonary septal defect. An anatomic review. Ped Cardiol 1982;3:329–35.
75. Broy H. Die spontane Darstellung der A. radicularis magna im abdominellen Aortogramm. Fortschr Röntgenstr 1974;120: 550–60.
76. Brucher J. Origin of the ophthalmic artery from the middle meningeal artery. Radiology 1969;93:51–2.
77. Bruetmann ME, Fields WS. Persistent hypoglossal artery. Hypoglossica primitiva. Arch Neurol 1963;8:369–72.
78. Bryan RN, Drewyer RG, Gee W. Separate origins of the left internal and external carotid arteries from the aorta. Am J Roentgenol 1978;130:362–5.
79. Büttner J. Die Blutgefäße im menschlichen Thalamus. Acta Anat 1960;41:279–99.
80. Busch W. Die arterielle Versorgung der Nebennieren (Zugleich ein Beitrag zur Anatomie der Nierenarterien). I. Mitteilung. Virchows Arch Path Anat 1954;324:688–99.
81. Caix M. Variations topographiques des pédicules hépatiques et spléniques. Ass Anat (Nancy) 1959;47:231–42.
82. Caix M, Descottes B, Rousseau D, Grousseau D. The arterial vascularisation of the middle thoracic and lower esophagus. Anat Clin 1981;3:95–106.
83. Campbell RL, Dyken ML. Four cases of carotid-basilar anastomosis associated with central neuron system dysfunction. J Neurol Neurosurg Psychiatry 1961;24:250–3.
84. Carbonin C, Benedetti A, Curri D, Rubini L. A case of persistent hypoglossal artery. Neurochirurgia 1976;19:231–6.
85. Carpenter MB, Noback CR, Moss ML. The anterior choroidal artery. Its origins, distribution and variations. Arch Neurol Psychiatry 1954;71:714–22.
86. Castañeda AR. Pulmonary artery sling (editorial). Ann Thorac Surg 1980;28:210–1.
87. Castañeda AR, Indeglia RA, Varco RL. Anomalous origin of the left coronary artery from the pulmonary artery. Certain therapeutic considerations. Circulation 1966;33 (suppl I): 52–6.
88. Catry J, Bossaert H, Collen D, Verstraete M. De klinische betekenis van de verbindingen tussen abdominale vaatgebieden. Acta Clin Belg 1969;24:299–315.
89. Cauldwell EW, Sieckert RG, Lininger RE, Anson BJ. The bronchial arteries. An anatomic study of 150 human cadavers. Surg Gynec Obstet 1948;86:395–412.
90. Celoria GC, Patton RB. Congenital absence of the aortic arch. Am Heart J 1959;58: 407–13.
91. Chadzypanagiotis D, Amerski L. A rare case of the celiac trunk anomaly. Folia Morphol (Warszawa) 1978;37:401–5.
92. Chakravorty BG. Arterial supply of the cervical spinal cord (with special reference to the radicular arteries). Anat Rec 1971; 170:311–30.
93. Chanamirjan A. Versuch eines systematischen Studiums der Variationen der Wirbelarterie. Anat Anz 1929;68:163–78.
94. Chang SC, Chen EY, Sung WL. Observations on the deep and circumflex femoral arteries in Chinese. Acta Anat Sinica 1965;8:282–92.
95. Charles CM, Penn L, Holden HF, Miller RA, Alvis EB. The origin of the deep brachial artery in American white and in American negro males. Anat Rec 1931; 50:299–302.
96. Chen E, Chang S, Lishing W. Observation of the superior mesenteric artery in Chinese adults. Acta Anat Sinica 1966;9:138–46.
97. Chérigié E, Mellière D, Bennet J, Doyon D, Chenard JC. Anatomie radiologique de la vascularisation du pancréas. J Radiol Electr 1967;48:346–52.
98. Chevrel JP, Guéraud JP. Arteries of the terminal ileum. Diaphanization study and surgical applications. Anat Clin 1978; 1:95–108.

99. Chiras J, Launay M, Gaston A, Bories J. Thoracic vertebral artery. An anomaly of the vertebral artery. Neuroradiology 1982;24:67–70.
100. Chiras J, Merland JJ. L'artère dorso-spinale. Une variante anatomique méconnue. Son importance en angiographie médullaire. J Neuroradiol 1979;6:93–100.
101. Chrzanowska G, Mierzwa J. Variation of the cerebral arterial circle of Willis in man. Folia Morphol (Warszawa) 1979;38:365–73.
102. Clausen HJ. An unusual variation in origin of the hepatic and splenic arteries. Anat Rec 1955;123:315–40.
103. Coleman SS, Anson BJ. Arterial patterns in the hand based upon a study of 650 specimens. Surg Gynecol Obstet 1961; 113:409–24.
104. Collins-Nakai RL, Dick M, Parisi-Buckley L, Fyler DC, Castañeda AR. Interrupted aortic arch in infancy. J Pediatr 1976; 88:959–62.
105. Combe J, Gallinet D, Weill F, Milleret P. A study of thirty right hepatic arteries. Int Surg 1976;61:112–6.
106. Copleman B. Anomalous right subclavian artery. Am J Roentgenol 1945;54:270–5.
107. Coulouma B. Resultats de cent quatre observations sur la disposition des branches de l'axillaire. Bull Ass Anat (Nancy) 1934; 29:193–9.
108. Crelin ES. An unusual anomalous blood vessel connecting the renal and internal spermatic arteries. Anat Rec 1948; 102:205–11.
109. Crudeli R. Observations sur les aspects anatomiques de l'artère communicante antérieure. Neurochirurgie 1960;6:29–37.
110. D'Cruz IA, Cantez T, Namin EP, Licata R, Hastreiter AR. Right-sided aorta. Part II. Right aortic arch, right descending aorta and associated anomalies. Br Heart J 1966;28:725–39.
111. Cudkowicz L, Armstrong JB. Observations on normal anatomy of the bronchial arteries. Thorax 1951;6:343–58.
112. Czerwinski F. Variability of the course of external carotid artery and its rami in man in the light of anatomical and radiological studies. Folia Morphol (Warszawa) 1981;40:449–53.
113. Daseler EH, Anson BJ. Surgical anatomy of the subclavian artery and its branches. Surg Gynec Obstet 1959;108:149–74.
114. Daseler EH, Anson BJ, Hambley WC, Reimann AF. The cystic artery and constituents of the hepatic pedicle. Surg Gynecol Obstet 1947;85:47–63.
115. Davies ER, Sutton D. Hypertension and multiple renal arteries. Am Heart J 1966;71:285–6.
116. Davis GD, Kincaid OW, Hunt JC. Roentgenologic evaluation of multiple renal arteries. Am J Roentgenol 1963;90:585–92.
117. Davis JS, Lie JT. Anomalous origin of a single coronary artery from the innominate artery. Angiology 1977;28:775–8.
118. Decker K, Hipp E. Der basale Gefäßkranz. Morphologie und Angiographie. Anat Anz 1958;105:100–16.
119. Delmas A, Gonzalez M. Angioarchitecture de l'oesophage. Bull Acad Nat Med 1971;155:621–6.
120. Derrick JR, Cleveland BR, Frazier J. Intramural and anomalous coronary arteries. Surgery 1963;53:315–8.
121. Derrick JR, Tyson KRT. The surgical significance of abberrant renal arteries in relation to systemic hypertension. Circulation 1961;24:1192–6.
122. Didio LJA, Christoforidis AJ, Chandnani PC. Posterior gastric artery and its significance as seen in angiograms. Am J Surg 1980;139:333–7.
123. Diemel H, Rau G, Schnitz-Dräger HG. Die Riolansche Kollaterale. Fortschr Geb Röntgenstr Nuklearmed 1964;101:253–64.
124. Diemel H, Schmitz-Dräger HG. Intraabdominelle Kollateralbahnen bei Verschlußkrankheiten der Eingeweidearterien. Fortschr Geb Röntgenstr Nuklearmed 1965;103:652–64.
125. Dimtza A. Die Arteriographie der Hand. Radiol Clin 1956;25:305–20.
126. Doerr W. Morphogenese und Korrelation chirurgisch wichtiger angeborener Herzfehler. Erg Chir Orthop 1950;36:1–92.
127. Dolton EG, Jones HE. Congenital anomalies of the aortic arch. Lancet 1952;262:537–40.
128. Dominguez R, Oh KS, Dorst JP, Young LW. Left aortic arch with right descending aorta. Am J Roentgenol 1978;130:917–20.
129. Domisse GF. The blood supply of the spinal cord. J Bone Joint Surg 1974;56:225–35.
130. Domisse GF. The arteries, arterioles, and capillaries of the spinal cord. Surgical guidelines in the prevention of postoperative paraplegia. Ann R Coll Surg 1980;62:369–76.
131. Donaldson RM, Thornton A, Raphael MJ, Sturridge MF, Emanuael RW. Anomalous origin of the left anterior descending coronary artery from the pulmonary trunk. Eur J Cardiol 1979;10:295–300.
132. Donovan DL, Sharp WV. Persistent sciatic artery: two case reports with emphasis on embryologic development. Surgery 1984;95:363–6.

133. Doppman JL. Spinal angiography. In: Abrams HL, ed. Abrams angiography, vascular and interventional radiology, Vol. 1. 3rd ed. Boston: Little, Brown, 1983:315–35.
134. Doppman JL, diChiro G, Glancy DL. Collateral circulation through dilated spinal cord arteries in aortic coarctation and extraspinal arteriovenous shunts. An arteriographic study. Clin Radiol 1969;20:192–7.
135. Doppman JL, diChiro G, Ommaya AK. Selective arteriography of the spinal cord. St. Louis: Warren Green, 1969:3–17.
136. Doscher W. Viswanathan B, Stein T, Margolis IB. Hemodynamic assessment of the circulation in 200 normal hands. Ann Surg 1983;198:776–9.
137. Doty DB, Chandramouli B, Schieken RE, Lauer RM, Ehrenhaft JL. Anomalous origin of the left coronary artery from the right pulmonary artery. J Thorac Cardiovasc Surg 1976;71:787–91.
138. Düx A, Hilger HH, Schaede A, Thurn P. Zur Koronarographie. Koronararterienbefunde im selektiven Aorten- und Laevokardiogramm bei angeborenen und erworbenen Herzfehlern. Fortschr Geb Röntgenstr Nuklearmed 1961;95:1–23.
139. Dunker RO, Harris AB. Surgical anatomy of the proximal anterior cerebral artery. J Neurosurg 1976;44:359–67.
140. Dunn JM, Gordon I, Chrispin AR, Leval MR de, Stark J. Early and late results of surgical correction of pulmonary artery sling. Ann Thorac Surg 1979;28:230–8.
141. Eckmann I, Krahn V. Untersuchung zur Häufigkeit verschiedener Ursprungsstellen der Arteria gastrica dextra. Anat Anz 1984;155:65–70.
142. Edsman G. Angionephrography and suprarenal angiography. Acta Radiol 1957;155 (suppl):5–141.
143. Edwards EA. Anatomy of the small arteries of the foot and toes. Acta Anat 1960;41:81–96.
144. Edwards EA. Organization of the small arteries of the hand and digits. Am J Surg 1960;99:837–46.
145. Edwards JE. Anomalies of the derivatives of the aortic arch system. Med Clin N Am 1948;32:925–44.
146. Edwards JE. Pathology of anomalies of thoracic aorta. Am J Clin Pathol 1953; 23:1240–58.
147. Edwards JE. Malformations of the aortic arch system manifested as "vascular rings". Lab Invest 1953;2:56–75.
148. Edwards JE. Congenital malformations of the heart and great vessels. In: Gould SE, ed. Pathology of the heart. 2nd ed. Springfield: Thomas, 1960:260–496.
149. Edwards JE. The direction of blood flow in coronary arteries arising from the pulmonary trunk. Circulation 1964;29:163–6.
150. Edwards JE. Anomalies of the aortic arch system. Birth Defects 1977;13:47–63.
151. Edwards JE, Dry TJ, Parker RL, Burchell HB, Wood EH, Bulbulian AH. An atlas of congenital anomalies of the heart and great vessels. Springfield: Thomas, 1954.
152. Edwards JE, McGoon DC. Absence of anatomic origin from heart of pulmonary arterial supply. Circulation 1973;17:393–8.
153. Ehler E, Hackenberg P. Gefäßabnormitäten im Retroperitonealsitus eines 68jährigen Mannes. Anat Anz 1959;107:375–83.
154. Eklöf O, Ekström G, Erikson BO, et al. Arterial anomalies causing compression of the trachea and/or the oesophagus. Acta Pediatr Scand 1971;60:81–9.
155. Ekström G, Sandblom P. Double aortic arch. Acta Chir Scand 1951;102:183–202.
156. El-Eishi HI, Ayoub SF, Adb-El-Khalek M. The arterial supply of the human stomach. Acta Anat 1973;86:565–80.
157. Ellis FH, Grindlay JH, Edwards JE. The bronchial arteries. Surgery 1951;30:810–26.
158. Ellis K, Seaman WB, Griffiths SP, Berdon WE, Baker DH. Some congenital anomalies of pulmonary arteries. Sem Roentgenol 1967;2:325–41.
159. Elze C. Ursprung der Arteria mesenterica superior aus der A. mesenterica inferior und der A. mesenterica inferior aus der A. mesenterica superior. Wien Klin Wochenschr 1962;74:109–10.
160. Engel HJ, Torres C, Page HL. Major variations of anatomical origin of the coronary arteries: an angiographic observation in 4,250 patients without associated congenital heart disease. Cathet Cardiovasc Diagn 1975;1:157–69.
161. Ennabli E. L'artère intercostale supérieure. Arch Anat Path 1966;14:A98–A103.
162. Ergin MA, Jayaram N, LaCorte M. Left aortic arch and right descending aorta: diagnostic and therapeutic implications of a rare type of vascular ring. Ann Thorac Surg 1981;31:82–5.
163. Éspalieu P, Cuilleret J, Charret P, Bsata A, Balique J-G. Étude radio-anatomique et micro-angiographique de la vascularisation systemique pulmonaire. Bull Assoc Anat 1982;66:441–50.
164. Faller A. Statistische Untersuchungen über Ursprung und Kaliber der Arteria transversa colli beim Menschen. Untersuchung

an 134 Halshälften. Arch Kreisl-Forsch 1952;18:161–7.
164a. Faller A, Schärer O. Über die Variabilität der Arteriae thyroideae. Acta Anat 1947; 4:119–22.
165. Faller J, Ungváry G. Die arterielle Segmentation der Niere. Zentralbl Chir 1962; 87:972–83.
166. Fauré C, Lefebvre J, Debrun G, Djindjian R. La vascularisation artérielle normale et pathologique du renflement lombaire de la moelle épinière chez l'enfant: l'artère d'Adamkiewicz. Ann Radiol 1967;10: 129–40.
167. Feigl W, Firbas W, Sinzinger H. Über das Vorkommen akzessorischer Leberarterien und ihre Genese sowie eine Diskussion ihrer Nomenklatur. Anat Anz 1973;134:139–71.
167a. Fernström I. Arteriography of the uterine artery. Acta Radiol (Diagn) (Stockh) 1955, Suppl 122, 1–128.
168. Fields WS. The significance of persistent trigeminal artery. Radiology 1968; 91: 1096–101.
169. Fields WS. Persistent hypoglossal artery in association with advanced atherosclerosis. Neurosurg Rev 1980;3:37–41.
170. Fine H, Keen EN. The arteries of the human kidney. J Anat 1966;100:881–94.
171. Finerty JC. Persistent ischiatic artery. Anat Rec 1947;98:587–95.
172. Finerty JC, Trotter M. Two unusual anomalies in the same cadaver (Negro male, age 85 years). I. Persistent right ischiadic artery. Anat Rec 1947;97:415.
173. Fischer W. Anatomisch-röntgenologische Untersuchungen über die Verteilung und Anastomosenbildung der visceralen und parietalen weiblichen Beckenarterien. Z Geburtsh Gynäk 1959;154:321–40.
174. Fishman NH, Bronstein MH, Berman W, et al. Surgical management of severe aortic coarctation and interrupted aortic arch in neonates. J Thorac Cardiovasc Surg 1976; 71:35–48.
175. Flint ER. Abnormalities in the right hepatic, cystic and gastro-duodenal arteries, and of the bile ducts. Br J Surg 1922;10:509–19.
176. Fontaine C, Libersa C, Chambon JP, Ribet M. La vascularisation artérielle de la queue du pancréas et les nécroses caudales pancréatiques après splénectomie. J Chir (Paris) 1980;117:677–82.
177. Fontaine R, Pietri J, Tongio J, Negreiros L. Étude angiographique des variations anatomiques des artères hépatiques basée sur 402 examens spécialisés. Angiology 1970; 21:110–3.
178. Foo D, Rossier AB. Anterior spinal artery syndrome and its natural history. Paraplegia 1983;21:1–10.
179. Fournial JF, Stanley P, Fouron JC, Guérin R, Davignon A. Les anomalies de l'arc aortique. À propos de 30 cas opérés chez l'enfant. Arch Mal Coeur 1975;68:485–95.
180. Fox M, Yalin R. Renal transplantation with multiple arteries. Br J Urol 1979;51:333–6.
181. Francke JP, Macke A, Clarisse J, Libersa JC, Dobbelaere P. The internal carotid arteries. Anat Clin 1982;3:243–61.
182. Francke JP, di Marino V, Pannier M, Argenson C, Libersa C. The vertebral arteries (arteria vertebralis). The V3 atlanto-axoidial and V4 intracranial segments-collaterals. Anat Clin 1981;2:229–42.
183. Franke H. Über Entwicklungs- und Lageanomalien der Aorta. Fortschr Geb Röntgenstr Nuklearmed 1950;73:267–80.
184. Fricke M, Haasner E. Seltene Variante eines Truncus coeliaco-mesentericus. Fortschr Geb Röntgenstr Nuklearmed 1974;2: 120:233–5.
185. Froehlich AL, Fujikura T. Significance of a single umbilical artery. Am J Obstet Gynec 1966;94:274–9.
186. Fujimoto Y, Suwa F, Kimura K. A case of the left superior thyroid artery arising from the left common carotid artery and the A. thyroidea ima. Okajimas Fol Anat Jap 1974;51:219–30.
187. Fulton WEM. The coronary arteries. Arteriography, microanatomy and pathogenesis of obliterative coronary artery disease. Springfield: Thomas, 1965.
188. Gans H. The anatomy of the intrahepatic structures and its repercussions on surgery. Amsterdam: Elsevier, 1955.
189. Garis CF de, Black IH, Riemenschneider EA. Patterns of aortic arch in American white and negro stocks, with comparative notes on certain other mammals. J Anat 1933;67:599–619.
190. Garis CF de, Swartley WB. The axillary artery in white and negro stocks. Am J Anat 1928; 41:353–97.
191. Garti IJ, Aygen MM. Left aortic arch with congenital isolation of the right subclavian artery. Pediatr. Radiol 1980;9:241–3.
192. Garti IJ, Aygen MM, Levy MJ. Double aortic arch anomalies: diagnosis by countercurrent right brachial arteriography. Am J Roentgenol 1979;133:251–6.
193. Garti IJ, Aygen NM, Vidne B. Type C double aortic arch. Double aortic arch with aberrant left subclavian artery. Cardiovasc Radiol 1978;1:143–5.

194. Garti IJ, Aygen MM, Vidne B, Levy MJ. Right aortic arch with mirror image branching causing vascular ring. A new classification of the right aortic arch patterns. Br J Radiol 1973;46:115–9.
195. Garti I, Meiraz D. Ectopic origin of main renal artery. Urology 1980;15:627–9.
196. Gasul BM, Loeffler E. Anomalous origin of the left coronary artery from the pulmonary artery (Bland-White-Garland-syndrome). Report of four cases. Pediatrics 1949;4:498–507.
197. Gelfand DW, Ott DJ. Anatomy and technique in evaluating the esophagus. Semin Roentgenol 1981;16:168–82.
198. Gerlach J, Jensen HP, Spuler H, Viehweger G. Die Persistenz der Arteria primitiva hypoglossica. Arch Psychiatry 1962;203: 164–72.
199. Geyer JR, Poutasse EF. Incidence of multiple renal arteries on aortography. Report of a series of 400 patients, 381 of whom had arterial hypertension. J Am Med Ass 1962; 182:120–5.
200. Gieseler HJ. Anatomische und experimentelle Untersuchungen zur Mechanik und Symptomatologie der traumatischen Milzrupturen. Langenbecks Arch Klin Chir 1965;309:340–67.
201. Giordani M, Pinna CD. L'origine delle arterie bronchiali (ricerca di anatomia chirurgica). Gazz Int Med Chir 1956;61:2733–51.
202. Gisel A. Genese und chirurgische Bedeutung einer seltenen Varietät im Halsteil der Arteria vertebralis nebst Bemerkungen zur Nomenklatur der segmentalen Arterien. Gegenbaurs Morph Jahrb 1959;100:81–94.
202 a. Gmelin E. Die Pankreasarterien und ihre Variationen in anatomischen und angiographischen Untersuchungen. Mediz. Dissertation, Techn. Univ. München, 1977.
203. Gmelin E, Weiss HD, Swoboda G. Seltene arterielle Variationen bei der Oberbauch- und Pankreasangiographie. Fortschr Geb Röntgenstr Nuklearmed 1978;129:77–82.
204. Godtfredsen J, Wennevold A, Efsen F, Lauridsen P. Natural history of vascular ring with clinical manifestations. A follow-up study of eleven unoperated cases. Scand J Thorac Cardiovasc Surg 1977;11:75–7.
205. Goerttler K. Die Mißbildungen des Herzens und der großen Gefäße. In: Bargmann W, Doerr W, eds. Das Herz des Menschen, Band I. Stuttgart: Thieme, 1963:422–601.
206. Goldstein JD, Rabinovitch M, van Praagh R, Reid L. Unusual vascular anomalies causing persistent pulmonary hypertension in a newborn. Am J Cardiol 1979;43:962–8.
207. Gomez-Oliveros L. Anatomia de la circulación cerebral. Los troncos arteriales del cuello. In: Accidentes cerebrovasculares agudos. Madrid 1968.
208. Gordon DH, Martin EC, Kim YH, Kutcher R. Accessory blood supply to the liver from the dorsal pancreatic artery: an unusual anatomic variant. Cardiovasc Radiol 1978;1:199–201.
209. Gouley BA. Anomalous left coronary artery arising from the pulmonary artery (adult type). Am Heart J 1950;40:630–7.
210. Grand W. Microsurgical anatomy of the proximal middle cerebral artery and the internal carotid artery bifurcation. 1980;7:215–8.
211. Graves FT. The anatomy of the intrarenal arteries and its application to segmental resection of the kidney. Br J Surg 1954;42:132–9.
212. Graves FT. The aberrant renal artery (synopsis). Proc Roy Soc Med 1957;50:368–70.
213. Greeb J. Congenital anomalies of the iliofemoral artery. J Cardiovasc Surg 1977; 18:317–23.
214. Gremigni D. Sulla origine di alcuni rami collaterali dell' arteria femorale. Z Anat Entwickl.-Gesch 1968;127:43–54.
215. Griffiths JD. Extramural and intramural blood supply of colon. Br Med J 1961;5222:323–6.
216. Grisoli J, Piganiol G, Sedan R. L'artère cérébrale communicante antérieure. Bull Assoc Anat (Nancy) 1958;99:349–55.
217. Grover FL, Norton JB, Webb GE, Trikle JK. Pulmonary sling. Case report and collective review. J Thorac Cardiovasc Surg 1975;69:295–300.
218. Grundmann R, Schoenemann B, Horsch S, Kallenberg A. Chirurgische Bedeutung der Varianten und Aneurysmen der Oberbauchgefäße. Langenbecks Arch Chir 1980; 353:35–46.
219. Di Guglielmo L, Guttadauro M. Anatomic variations in the coronary arteries. An arteriographic study in living subjects. Acta Radiol Diagn (Stockh) 1954;41:393–416.
220. Di Guglielmo L, Guttadauro M. A roentgenologic study of the coronary arteries in the living. Acta Radiol Diagn (Stockh) 1972; Suppl. 97, 1–82.
221. Guinto FC, Garrabrant EC, Radcliffe WB. Radiology of the persistent stapedial artery. Radiology 1972;105:365–9.
222. Gupta CD, Gupta SC, Arora AK, Singh PJ. Vascular segments of the human spleen. J Anat 1976;121:613–6.

223. Gupta SC, Gupta CD, Gupta SB. Intrahepatic supply patterns in cases of double hepatic arteries – a study by corrosion casts. Anat Anz 1979; 146: 166–70.
224. Guyader A le, Kerech K, Delteil C, Laigo M. La vascularisation artérielle du colon du nouveau-né Africain de Côte d'Ivoire. CR Ass Anat 1966; 132: 653–62.
225. Gwyn DG, Skilton JS. A rare variation of the inferior mesenteric artery in man. Anat Rec 1966; 156: 235–8.
226. Habighorst LU. Die Gefäße der Gallenblase im postmortalem Angiogramm. Fortschr Röntgenstr 1965; 103: 63–70.
227. Habighorst LU, Albers P, Zeitler E. Die Gefäße der Gallenblase im postmortalen Angiogramm. Fortschr Geb Röntgenstr 1965; 103: 63–70.
228. Habighorst LU, Kössling FK, Albers P. Die perirenalen Arterien und der Kollateralkreislauf der Niere im postmortalen Angiogramm. Fortschr Geb Röntgenstr Nuklearmed 1966; 105: 35–48.
229. Hackensellner HA. Koronaranomalien unter 1000 auslesefrei untersuchten Herzen. Anat Anz 1954; 101: 123–30.
230. Hackensellner HA. Arteria subclavia lusoria bei einem 14,5 mm langen menschlichen Embryo. Anat Anz 1955; 102: 204–9.
231. Hackensellner HA. Über akzessorische, von der Arteria pulmonalis abgehende Herzgefäße und ihre Bedeutung für das Verständnis der formalen Genese des Ursprungs einer oder beider Koronararterien von der Lungenschlagader. Wien Klin Wochenschr 1956; 68: 35.
232. Hackensellner HA. Akzessorische Kranzgefäßanlagen der Arteria pulmonalis unter 63 menschlichen Embryonenserien mit einer größten Länge von 12–36 mm. Z Mikr Anat Forsch 1956; 62: 153–64.
233. Hackensellner HA, Stur D. Korrelation und Kombination von Anomalien und Varietäten des Herzens und der herznahen Gefäße unter 1234 Obduktionen. Zentralbl Pathol 1956; 96: 188–92.
234. Hafferl A. Das Arteriensystem. In: Bolk L, Göppert E, Kallius E, Lubosch W, eds. Handbuch der vergleichenden Anatomie der Wirbeltiere. Bd. VI. Amsterdam: Asher, 1967.
235. Hahn von Dorsche H, Dorn A. Varietäten im Bereich der A. coeliaca und an den Nierengefäßen. Anat Anz 1966; 118: 297–304.
236. Hall CL, Sansom JR, Obeid M et al. Agonal phase, ischaemic times and renal vascular abnormalities and outcome of cadaver kidney transplants. Br Med J 1975; 3: 667–70.
237. Hallman GL, Cooley DA, Singer DB. Congenital anomalies of the coronary arteries: anatomy, pathology and surgical treatment. Surgery 1966; 59: 133–44.
238. Hanai H, Ryümon A, Kamakura K, Kaneko S. An anomalous case of a subclavia dextra rising from arcus aortae as the last branch. Okajimas Folia Anat Jap 1963; 39: 31–5.
239. Handa J, Kikuchi H, Handa H. Persistierende karotido-basiläre Anastomose: Die A. primitiva hypoglossica. Fortschr Röntgenstr 1967; 107: 421–2.
240. Handa J, Seta K, Handa H. Die akzessorische A. cerebri media. Fortschr Röntgenstr 1968; 108: 539–41.
241. Hannoun L, Le Breton C, Bors V, Helenon C, Bigot JM, Parc R. Radiological anatomy of the right gastroepiploic artery. Anat Clin 1984; 5: 265–71.
242. Hanssen HH. Surgical anatomy of the blood supply of the colon. Langenbecks Arch Chir 1975; 340: 63–74.
243. Harley HRS. The development and anomalies of the aortic arch and its branches with the report of a case of right cervical aortic arch and intrathoracic vascular ring. Br J Surg 1959; 46: 561–73.
244. Harrison CR, Luttrell C. Persistent carotid-basilar anastomosis. Three arteriographically demonstrated cases with one anatomical specimen. J Neurosurg 1953; 10: 205–15.
245. Harrison LH, Flye MW, Seigler HF. Incidence of anatomical variants in renal vasculature in the presence of normal renal function. Ann Surg 1978; 188: 83–9.
246. Harrison RG, McGregor GA. Anomalous origin and branching of the testicular arteries. Anat Rec 1957; 129: 401–5.
247. Hassan ST, Khan MA. Abnormal origin of the right subclavian artery and common origin of the right and the left common carotid arteries. Medicus Karachi 1966; 31: 297–300.
248. Hastreiter AR, D'Cruz IA, Cantez T. Right-sided aorta. Part I. Occurrence of right aortic arch in various types of congenital heart disease. Br. Heart J 1966; 28: 722–4.
249. Haug H, Merkle U. Abnormer Ursprung und Verlauf der A. subclavia dextra. Anat Anz 1965; 117: 319–21.
250. Haughton VM, Fellows KE, Rosenbaum AE. The cervical aortic arches. Pediatr Radiol 1975; 114: 675–81.
251. Hazlett JW. The superficial ulnar artery with reference to accidental intra-arterial injection. Can Med Ass J 1949; 61: 289–93.
252. Hellekant C, Tylén U. Leftside intercostobronchial trunk: a rare anomaly. Am J Roentgenol 1980; 134: 590–1.

253. Hellerer O, Schmidt GP, Aigner R. Die A. thyreoidea ima – eine chirurgisch bedeutsame Variation. Chir Prax 1978;24:413–7.
254. Henschen C. Die chirurgische Anatomie der Milzgefäße. Schweiz Med Wochenschr 1928;9:164–77.
255. Hentschel M. (42.) Pankreas-Anatomie. Eine Studie aus chirurgischer und röntgenologischer Sicht. Langenbecks Arch Klin Chir 1965;313:233–42.
256. Hepp J, Hernandez C, Moreaux J, Bismuth H. L'artériographie dans les affections chirurgicales du foie, du pancréas et de la rate. Paris: Masson 1968.
257. Hewitt RL, Brewer PL, Drapanas T. Aortic arch anomalies. J Thorac Cardiovasc Surg 1970;60:746–53.
258. Hinck VC, Gordy PD. Persistent primitive trigeminal artery. One type of persistent carotid-basilar anastomosis. Radiology 1964;83:41–5.
259. Ho SY, Wilcox BR, Anderseon RH, Lincoln JCR. Interrupted aortic arch – anatomical features of surgical significance. Thorac Cardiovasc Surg 1983;31:199–205.
260. Hochstetter, A.v. Eine einzigartig abnorme Anordnung der Arterien des Gesichts. Anat Anz 1963;113:221–3.
261. Hodes P, Campoy F, Riggs HE, Ely P. Cerebral angiography: fundamentals in anatomy and physiology. Am J Roentgenol 1953; 70:61–82.
262. Hoganson G, McPherson E, Piper P, Gilbert EF. Single coronary artery arising anomalously from the pulmonary trunk. Arch Pathol Lab Med 1983;107:199–201.
263. Hollinshead WH. Some variations and anomalies of the vascular system in the abdomen. Surg Clin North Am 1955;35:1123.
264. van der Horst RL, Fisher EA, DuBrow IW, Hastreiter AR. Right aortic arch, right patent ductus arteriosus and mirror-image branching of the brachiocephalic vessels. Cardiovasc Radiol 1978;1:147–9.
265. Househam KC, Human DG, Fraser CB, Joffe HS. Anomalous left coronary artery from the pulmonary artery – a therapeutic dilemma. A report on 12 cases. S Afr Med J 1983;63:325–7.
266. Huber G, Piepgras U. Ursprung der linken Arteria vertebralis aus der linken Arteria carotis externa. Fortschr Geb Roentgenstr Nuklearmed 1976;125:63–6.
267. Huber JF. The arterial network supplying the dorsum of the foot. Anat Rec 1941;80:373–91.
268. Huber P. Die A. trigemina primitiva. Betrachtungen zur klinischen Bedeutung der carotido-basilären Anastomose. Dtsch Z Nervenheilk 1961;181:612–33.
269. Huelke DF. Variation in the origins of the branches of the axillary artery. Anat Rec 1959;135:33–41.
270. Huu N, Tam NTT, Minh NK. Artère gastroduodénale naissant de la mésentérique supérieure. Bull Asso Anat (Nancy) 1978; 60:779–86.
271. Idbeis B, Levinsky L, Srinivasan V, Subramanian S. Vascular rings: management and a proposed nomenclature. Ann Thorac Surg 1981;31:255–8.
272. Immagoulou A, Anderson RC, Moller JH. Interruption of the aortic arch. Clinical features in 20 patients. Chest 1972;61:276–82.
273. Ionescu M, Mihail N, Ionescu C. Blutversorgung der normalen Niere des Menschen durch mehrfache Nierenschlagadern Anat Anz 1962;111:398–406.
274. Ishihara Y, Fukuda R, Awaya Y, et al. Anomalous systemic arterial supply to the basal segments of the lung presenting with a murmur. Eur J Pediatr. 1979;131:125–31.
275. Izukawa T, Scott ME, Durran F, Moes CAF. Persistent left fifth aortic arch in man. Report of two cases. Br Heart J 1973;35: 1190–5.
276. Jackson IJ, Garza-Mercado R. Persistent carotid-basilar artery anastomosis: occasionally a possible cause of tic douloureux. Angiology 1960;11:103–7.
277. Jaffe RB. Complete interruption of the aortic arch. I. Characteristic radiographic findings in 21 patients. Circulation 1975; 52:714–21.
278. Jain KK. Some observations on the anatomy of the middle cerebral artery. Can J Surg 1964;7:134–9.
279. James TN. Anatomy of the coronary arteries. New York: Hoeber, 1961.
280. James TN. Anatomy of the coronary arteries and veins. In: Hurst JW, Logue RB, eds. The heart. Arteries and veins. New York, Toronto, Sidney, London: McGraw-Hill, 1966: 628–41.
281. James TN, Burch GE. Blood supply of the human interventricular septum. Circulation 1958;17:391–6.
281 a. Jellinger K. Zur Orthologie und Pathologie der Rückenmarksdurchblutung. Wien, New York: Springer, 1966.
282. Juskiewenski S, Vaysse P. Arterial vascularization of the testes and surgery for undescended testicles (testicular ectopia) Anat Clin 1978;1:127–34.
283. Kadanoff D, Balkansky G. Zwei Fälle mit seltenen Varietäten der Arterien der oberen Extremitäten. Anat Anz 1966; 118:289–96.

284. Kägi J. Beitrag zur Topographie der Arteria transversa colli. Eine Untersuchung an 134 Halshälften. Anat Anz 1959;107:168–86.
285. Kafrouni G, Khan AH, Wolfsen JL. Single right coronary artery: clinical and angiographic findings with surgical management. Ann Thorac Surg 1981;32:80–4.
286. Kahn PC. Selective angiography of the inferior phrenic arteries. Radiology 1967;88:1–8.
287. Kahn P, Abrams HL. Inferior mesenteric arterial patterns. Radiology 1964;82:429–42.
288. Kamienski RW, Barnes RW. Critique of the Allen test for continuity of the palmar arch assessed by Doppler ultrasound. Surg Gynecol Obstet 1976;142:861–4.
289. Kamina P, Koumare AK, Rideau Y. Contribution à l'étude des variations de l'origine de l'artère rénale (à propos de 800 artériographies). Arch Anat Pathol (Paris) 1975;23:287–90.
289a. Karlsson S, Jonsson K. Angiography of the ovarian artery in adnexal lesions. Acta Radiol (Diagn) (Stockh) 1980;21:739–46.
290. Kasai T, Chiba S. Macroscopic anatomy of the bronchial arteries. Anat Anz 1979;145:166–81.
291. Katritsis E, Parashos A, Papadopoulos N. Arterial segmentation of the human spleen by post-mortem angiograms and corrosion-casts. Angiology 1982;33:720–7.
292. Keane JF, Maltz D, Bernhard WF, Corwin RD, Nadas AS. Anomalous origin of one pulmonary artery from the ascending aorta. Diagnostic, physiological and surgical considerations. Circulation 1974;50:588–94.
293. Keen JA. A study of the arterial variations in the limbs, with a special reference to symmetry of vascular patterns. Am J Anat 1961;108:245–61.
294. Kelemann G. Arteria stapedia, in bilateral persistence. Arch Otolaryngol 1958;67:668–77.
295. Keller HL. Varianten der Arteria carotis interna, der Arteria meningea media und der Arteria ophthalmica im Karotisangiogramm. Fortschr Röntgenstr 1961;95:472–82.
296. Keren A, Tzivoni D, Granot C, Yalon S, Stern S. Origin of both coronary arteries from the same sinus of valsalva: report of two cases. Cathet Cardiovasc Diagn 1982;8:191–9.
297. Kido KD, Baker RA, Rumbaugh CL. Normal cerebral vascular anatomy. In: Abrams HL, ed. Abrams angiography, Vol. 1, 3rd ed. Boston: Little Brown Co., 1983;231–70.
298. Kimbiris D, Iskandrian AS, Segal BL, Bemis CE. Anomalous aortic origin of coronary arteries. Circulation 1978;58:606–15.
299. Kirgis HD, Llewellyn RC, Peebles EM. Functional trifurcation of the internal carotid artery and its potential clinical significance. J Neurosurg 1960;17:1062–72.
300. Kirks DR, Kane PE, Free EA, Taybi H. Systemic arterial supply to normal basilar segments of the left lower lobe. Am J Roentgenol 1976;126:817–21.
301. Klinkhammer AC. Aberrant right subclavian artery. Clinical and roentgenologic aspects. Am J Roentgenol Radium Ther Nucl Med 1966;97:438–46.
302. Koehler RE, Korobkin M, Lewis F. Arteriographic demonstration of collateral arterial supply of the liver after hepatic artery ligation. Radiology 1975;117:49–54.
303. Kollmannsberger A. Vergleichende Studien über die A. Heubneri und die Aa. chorioideae an anatomischen Präparaten, an Korrosionspräparaten und Angiogrammen der Hirngefäße. Morphol Jahrb 1961;102:180–99.
304. Koritke JG, Gillet JY, Pietri J. Les artères de la trompe utérine chez la femme. Arch Anat Histol Embryol Norm Exp 1967;50:47–70.
305. Korns ME. Single coronary artery. Arch Path Chir 1964;78;169–71.
306. Kottmann UR, Schniepp KE. Rechtslage des Aortenbogens mit Atresie des linken Truncus brachiocephalicus. Ein Beitrag zu den Anomalien des Aortenbogens. Thoraxchir 1977;25:123–5.
307. Kou S, Takemura A, Irifune T, Irifune T. A case of the splenic artery arising from the superior mesenteric artery. Okajimas Folia Anat Jpn 1980;57:241–50.
308. Krasemann PH. Doppelte Arteria femoralis. Fortschr Geb Röntgenstr Nuklearmed 1972;117:220–2.
309. Krayenbühl H, Yasargil MG. Die zerebrale Angiographie, 2. Auflage. Stuttgart: Thieme, 1965.
310. Križan Z. Beiträge zur deskriptiven und topographischen Anatomie der A. maxillaris. Acta Anat 1960;41:319–33.
311. Križan Z. Über die fraglichen Korrelationen und über die Entwicklung einiger Typen der A. maxillaris. Acta Anat 1960;42:71–87.
312. Krmpotic-Nemanic J. Anatomie, Variationen und Mißbildungen der Gefäße im Kopf- und Halsbereich. Arch Oto Rhino Laryngol 1978;219:1–91.
313. Kubik S. Bau und Blutversorgung der bronchopulmonalen Segmente und ihre Varian-

ten. Ergebn Anat Entwickl Gesch 1962;36:112–270.
314. Kučera V, Fišer B, Tůma S, Hučin B. Unilateral absence of pulmonary artery. A report on 19 selected clinical cases. Thorac Cardiovasc Surg 1982;30:152–8.
315. Kukwa A, Zbrodowski A. Rzadki przypadek odejścia tetnicy tarczowej górney od tetnicy szyjnej wspólnej lewej. Folia Morphol (Warszawa) 1966;25:641–3.
316. Kurrat HJ, Hesse M. Doppelbildung der Arteria testicularis sinistra mit Ursprungs- und Verlaufsvarianten. – Eine entwicklungsgeschichtliche Betrachtung. Anat Anz 1979;145:303–7.
317. Lacoste P, Dalle M. Anomalie artérielle de l'avant-bras et de la main. Essai d'interprétation embryologique. Comp' Rend Ass Anat 1966;132:563–70.
318. Laffont J, Fustec R, Goudote E, Conty CR, Fournier JP. Considérations sur les systèmes artériels mésentériques chez l'Africain. CR Ass Anat 1966;132:577–80.
319. Lahl R, Pellicioni R, Ruser I, Lutter B. Zur klinischen Bedeutung der karotido-basilären Anastomose unter besonderer Berücksichtigung der A. primitiva trigemina. Psychiatr Neurol Med Psychol (Leipzig) 1982;34:469–80.
320. Lam CR, Kabbani S, Arciniegas E. Symptomatic anomalies of the aortic arch. Surg Gynecol Obstet 1978;147:673–81.
321. Lamarque JL, Senac JP, Bruel JM, Gau A, Toulza A. Anatomie radiologique de l'artère hépatique. Ann Radiol 1976;19:316–8.
322. Lang D, Hofstetter R, Müßig G, von Bernuth G. Kompression von Trachea und Oesophagus durch abnormen Verlauf der linken Pulmonalarterie. Klin Paediatr 1975; 187:546–50.
322a. Lang J. Kopf, Gehirn- und Augenschädel. In: Lanz T, Wachsmuth W, eds. Praktische Anatomie, vol. 1, part 1B. Berlin, Heidelberg: Springer, 1979.
322b. Lang J. Klinische Anatomie des Kopfes. Berlin, Heidelberg: Springer 1981.
323. Lang J, Baldauf R. Beitrag zur Gefäßversorgung des Rückenmarks. Gegenbaurs Morph Jahrb 1983;129:57–95.
324. Lang J, Dehling U. A. cerebri media. Abgangzonen und Weiten ihrer Rami corticales. Acta Anat 1980;108:419–29.
325. Lang J, Häckel HR. Neue Befunde zum Verlauf der A. cerebri anterior (pars postcommunicalis) zu den Abgangszonen und Weiten ihrer Rami corticales. Acta Anat 1980;108:498–509.
326. Lang J, Heiler E. Anatomisch-klinische Befunde zur A. pharyngea ascendens. Anat Anz 1984;156:177–207.
327. Lang J, Kollmannsberger A. Beitrag zur Anatomie der Kleinhirnarterien. Gegenbaurs Morphol Jahrbuch 1961;102:170–9.
328. Lange-Cosack H. Anatomie und Klinik der Gefäßmißbildungen des Gehirns und seiner Häute. In: Handbuch der Neurochirurgie, Band IV/2. Berlin, Heidelberg, New York: Springer, 1966:1–145, 364–377.
329. Langlois J, Binet JP, Planche C, et al. La compression de la trachée par le tronc artériel brachio-céphalique chez le nourrisson. À propos d'une expérience chirurgicale de trente-quartre cas. Ann Chir 1982;36: 157–63.
330. Lanz T, Wachsmuth W. Praktische Anatomie, Vol 1, part 4 Bein und Statik. Berlin: Springer, 1938.
331. Lanz T, Wachsmuth W. Praktische Anatomie Vol. I, Part 2. Hals. Berlin, Göttingen, Heidelberg: Springer, 1955.
332. Lanz T, Wachsmuth W. Praktische Anatomie, Vol 1, part 3 Arm. 2nd ed. Berlin: Springer 1959.
333. Lasjaunias PL. Craniofacial and upper cervical arteries. Functional, clinical and angiographic aspects. Baltimore, London: Williams & Wilkins 1981.
334. Lasjaunias P, Moret J, Manelfe C, Théron J, Hasso T, Seeger J. Arterial anomalies at the base of the skull. Neuroradiology 1977;13:267–72.
335. Lasjaunias P, Théron J, Moret J. The occipital artery. Anatomy, normal arteriographic aspects, embryological significance. Neuroradiology 1978;15:31–7.
336. Lasker GW, Opdyke DL, Miller H. The position of the internal maxillary artery and its questionable relation to the cephalic index. Anat Rec 1951;109:119–26.
337. Latarjet M, Juttin P. Données nouvelles sur la circulation dans les artères bronchiques. Le Poumon 1951;7:35–50.
338. Latour H, Méry D, Ferrière M, Baissus C, Grolleau-Raoux R, Puech P. Les hypoplasies parenchymateuses et les dysplasies artérielles des reins dans l'hypertension artérielle de l'adulte. Donées de l'artériographie rénale sélective. Arch Mal Coeur 1975;68:1149–63.
339. Lavasseur JC, Couinaud C. Étude de la distribution des artères gastriques. J Chir (Paris) 1968;95:161–172.
340. Lawrence TYK, Stiles QR. Persistent fifth aortic arch in man. Am J Dis Child 1975;129:1229–31.

341. Lazorthes G, Gouaze A, Santini JJ, Salamon G. The arterial circle of the brain (circulus arteriosus cerebri). Anatomia Clinica 1979;1:241–57.
342. Lazorthes G, Gouaze A, Zadeh JO, Santini JJ, Lazorthes Y, Burdin P. Arterial vascularization of the spinal cord. Recent studies of the anastomotic substitution pathways. J Neurosurg 1971;35:253–62.
343. Lazorthes G, Zadeh JO. Précisions sur la vascularisation de la moelle dorsale. Bull Ass Anat (Nancy) 1982;66:477–80.
344. Leguerrier A, Calmat A, Honnart F, Cabrol C. Variations anatomiques des orifices coronariens aortiques. (À propos de 80 dissections) Bull Assoc Anat (Nancy) 1976;60:721–39.
345. Lemmi H, Daly RF, Kashgarian M, Argires J. A comparison of the level of bifurcation of the common carotid arteries in Negroes and Whites. Acta Anat 1968;71:22–7.
346. Lerberg DB, Ogden JA, Zuberbuhler JR, Bahnson HT. Anomalous origin of the right coronary artery from the pulmonary artery. Ann Thorac Surg 1979;27:87–94.
347. Levasseur JC, Couinaud C. Étude de la distribution des artères gastriques. Incidences chirurgicales (1). J Chir (Paris) 1968; 95:57–78.
348. Levasseur JC, Couinaud C. Étude de la distribution des artères gastriques. II. Incidences chirurgicales. J Chir (Paris) 1968; 95:161–76.
349. Levy M, Pauker M. Carpal tunnel syndrome due to thrombosed persisting median artery. A case report. Hand 1978;10:65–8.
350. Libersa C, Francke JP, Mauppin JM, Bailleul JP, Gamblin P. The arterial supply to the palm of the hand (arteriae palmae manus). Anat Clin 1982;33–45.
351. Liberthson RR, Dinsmore RE, Bharati S, et al. Aberrant coronary artery origin from the aorta. Diagnosis and clinical significance. Circulation 1974;50:774–9.
352. Liberthson RR, Dinsmore RE, Fallon JT. Aberrant coronary artery origin from the aorta. Report of 18 patients, review of literature and delineation of natural history and management. Circulation 1979;59:748–54.
352a. Lie TA. Congenital anomalies of the carotid arteries. Baltimore: Williams & Wilkins, 1968.
353. Lie TA. Congenital malformations of the carotid and vertebral arterial systems, including the persistent anastomoses. In: Vinken PJ, Bruyn GW, eds. Handbook of clinical neurology. Vol. 12. New York: Elsevier 1972:289–339.
354. Liebow AA. Patterns of origin and distribution of the major bronchial arteries in man. Am J Anat 1965;117:19–32.
355. Liechty JD, Shields TW, Anson BJ. Variations pertaining to aortic arches with their branches; with comments of surgically important types. Quart Bull Northwestern Univ Med School 1957;31:138–43.
356. Lierse W. Becken. In: Lanz T, Wachsmuth W, eds. Praktische Anatomie. Vol II, Part 8a. Berlin: Springer 1984.
357. Lippert H, Werner T. Arcus aortae dexter als Zufallsbefund. Med Klin 1969;64:382–7.
358. Lipshutz B. A composite study of the hypogastric artery and its branches. Ann Surg 1918;67:584–608.
359. Little JM, Zylstra PL, West J, May J. Circulatory patterns in the normal hand. Br J Surg 1973;60:652–5.
360. Ljungqvist A, Lagergren C. Normal intrarenal arterial pattern in adult and ageing human kidney. J Anat 1962;96:285–300.
361. Loetzke HH, Kleinau W. Gleichzeitiges Vorkommen der Aa. brachialis superficialis radialis und antebrachialis dorsalis superficialis sowie der Aufzweigungen. Anat Anz 1968;122:137–41.
362. Lohmann AHM, de Kort EJ. Eine Abnormität des Ursprunges und Verlaufes der Bronchialarterien der linken Lunge. Anat Anz 1963;112:12–8.
363. Longenecker CG, Reemtsma K, Chreech O. Surgical implications of single coronary artery. A review and two case reports. Am Heart J 1961;61:382–6.
364. Loue F, Cayotte JL, Borrelly J. Étude radioanatomique des artères bronchiques (à propos de 30 préparations analysables). Bull Assoc Anat 1976;60:741–5.
365. Lüdinghausen M von. Das Verteilungsmuster der Koronararterien und ihr Einbau in das Myokard. Ein Beitrag zur funktionellen Anatomie der Herzkranzgefäße. Dtsch Med Wochenschr 1975;100:2448–51.
366. Lüdinghausen M von, Deiler S. Präparatorische, röntgenologische und histologische Untersuchungen der Mesenterialarterien und ihrer Äste im Gebiet des Colon transversum und des Colon descendens im höheren Alter. Verh Anat Ges 1984;79: im Druck.
367. Lunderquist A. Arterial segmental supply of the liver. An angiographic study. Acta Radiol (Stockh) 1967;272 (suppl) 1–135.
368. Lurje A. On the topographical anatomy of the internal maxillary artery. Acta Anat 1947;2:219–31.
369. Luyendijk W. De arteriën van het ruggemerg. Ned Tijdschr Geneeskd 1981; 125:616–20.

370. Luyendijk W. The arteries of the spinal cord. The history of a paradigmal shift. Acta Neurochir 1982;61:25–41.
371. Macchi RJ, Fabregas RA, Chianelli HO, Bourdet JC, Lhez O, Stagnaro R. Anomalous communication of the left coronary artery with a peripheral branch of the right pulmonary artery. Chest 1976;69:565–8.
372. Mahoney EB, Manning JA. Congenital abnormalities of the aortic arch. Surgery 1964;55:1–14.
373. Malchiodi L, Ruberti U. Osservazioni anatomo-radiologiche sulle anomalie congenite delle arterie dell'arto inferiore. (Dall'analisi di 650 arteriografie femorali). Minerva Cardio-Angiol (Torino) 1957;5:297–9.
374. Manouguian S, Bjoernstad PG, Rastan H. Unterbrechung des Aortenbogens mit Fallot'scher Tetralogie (Bericht über eine einzeitige erfolgreiche Korrektur beider Mißbildungen. Zugleich eine Literaturübersicht). Thoraxchirurgie 1975;23:176–87.
375. Manz F, Conradi R, Neumann J, Wolf D. Angeborene hochgradige Trachealstenose durch anomalen Verlauf der Arteria pulmonalis sinistra. Monatsschr Kinderheilkd 1977;125:668–70.
376. Marano GD, Horton JA, Gabriele OF. Persistent embryologic vascular loop of the internal carotid, middle meningeal and ophthalmic arteries. Radiology 1981; 141:409–10.
377. Maraspin LE. A report of an anomalous bifurcation of the right axillary artery in man. Vasc Surg 1971;5:133–6.
378. Marchal G, Kint E, Nijssens M, Baert AL. Variability of the hepatic arterial anatomy: a sonographic demonstration. J Clin Ultrasound 1981;9:377–81.
379. Marchand P, Gilroy JC, Wilson VH. An anatomical study of the bronchial vascular system and its variations in disease. Thorax 1950;5:207–21.
380. Marongiu G, Dettori G, Madeddu G, et al. Studio dei rapporti fra nervo ricorrente ed arteria tiroidea inferiore. Chir Ital 1981;33:76–85.
381. Martin K. Ungewöhnlicher Verlauf einer Arteria suprarenalis. Acta Anat 1955; 24:48–50.
382. Martin MJ, Whisnant JP, Sayre GP. Occlusive disease in extracranial circulation. Arch Neurol 1960;3:530–8.
383. Martin R, Couppié G. Étude des variations du tronc coeliaque (d'après 229 pièves anatomiques). Arch Anat Path (Paris) 1960;8:27–30.
384. Mayo CW. Blood supply of the colon. Surgical considerations. Surg Clin North Am 1955;35:1117–22.
385. Mayschak DT, Flye MW. Treatment of the persistent sciatic artery. Ann Surg 1984;199:69–74.
386. McConwell EM. The arterial blood supply of the human hypophysis cerebri. Anat Rec 1953;115:175–203.
387. McCormack LJ, Cauldwell W, Anson BJ. Brachial and antebrachial arterial patterns. A study of 750 extremities. Surg Gynec Obstet 1953;96:43–54.
388. McCullough AW. Some anomalies of the cerebral arterial circle of Willis and related vessels. Anat Rec 1962;142:537–43.
389. McDowell DE, Grant MA, Gustafson RA. Single arterial trunk arising from the aortic arch. Circulation 1980;62:181–2.
390. McLaughlin RF. Bronchial artery distribution in various mammals and in humans. Am Rev Respir Dis 1983;128:S57–8.
391. McLellan GL, Morettin LB. Persistent sciatic artery: Clinical ,surgical and angiographic aspects. Arch Surg 1982;117:817–22.
392. McLennan JE, Rosenbaum AE, Haughton VM. Internal carotid origins of the middle meningeal artery. The ophthalmic-middle meningeal and stapedial-middle meningeal arteries. Neuroradiology 1974;7:265–75.
393. Meitner ER. Über zwei Fälle von Pulmonalis-Atresie mit abnormem Verhalten der Aa. bronchiales. Anat Anz 1958;105:148–53.
394. Mellière D. Variations des artères hépatiques et du carrefour pancréatique. J Chir (Paris) 1968;95:5–42.
395. Mellière D. Variations in the abdominal arteries applied to radical surgery and transplantation procedures. Rev Surg 1968; 25:305–13.
396. Mendez R, Mendez RG, Pabne JE, Silcott GR, Berne TV. Management of multiple renal arteries in renal transplantation. Urology 1974;3:409–12.
397. Mentha C. Malposition et sténose extrinsèque de l'artère poplitée par la compression musculo-tendineuse du jumeau interne. J Chir (Paris) 1966;91:489.
398. Merklin RJ. Arterial supply of the suprarenal gland. Anat Rec 1962;144:359–71.
399. Merklin RP, Michaels NA. The variant renal and suprarenal blood supply with data on the inferior phrenic, ureteral and gonadal arteries. J Int Coll Surg 1958;29:41–6.
399a. Michels NA. Blood supply and anatomy of the upper abdominal organs with a descriptive atlas. Philadelphia: Lippincott, 1955.
400. Michels NA. The ever varied blood supply of the liver and its collateral circulation. J Int Coll Surg 1957;27:1–17.
401. Michels NA. Newer anatomy of liver – variant blood supply and collateral circulation. J Am Med Ass 1960;172:125–32.

402. Michels NA. Renal circulation. Gross anatomy. In: Abramson DI, ed. Blood vessels and lymphatics. Philadelphia, London: Academic Press, 1962:390–4.
402 a. Michels NA, Siddharth P, Kornblith PL, Park WW. The variant blood supply of the descending colon, rectosigmoid and rectum based on 400 dissections. Its importance in regional dissections, a review of medical literature. Dis Colon Rectum 1965;8:251–78.
403. Michels N, Siddharth P, Kornblith P, Parke W. Routes of collateral circulation of the gastrointestinal tract as ascertained in a dissection of 500 bodies. Int Surg 1968; 49:8–28.
404. Mierzwa J, Chrzanowska G. Arteries of horseshoe kidney with fused lower poles. Folia morphol (Warszawa) 1978;37:381–8.
405. Miller RA. Observations upon the arrangement of the axillary artery and brachial plexus. Am J Anat 1939;64:143–63.
406. Milon J, Milon D, Boullier G, le Guerrier A, Lanchou G. L'artère gastro-duodénale et ses variations. À partir de cent artériographies. Bull Ass Anat (Nancy) 1978;62:461–3.
407. Mitterwallner F von. Variations-statistische Untersuchungen an den basalen Hirngefäßen. Acta Anat 1955;24:51–87.
408. Młynarczyk L, Woźniak W, Kiersz A. Varianten in Anzahl und im Verlauf der Nierenarterien. Anat Anz 1966;118:67–81.
409. Moberg A. Anastomoses between extracardiac vessels and coronary arteries. I. Via bronchial arteries. II. Via internal mammary arteries. Acta Radiol Diagn 1967;6:177–92, 263–72.
410. Moberg A. Anastomoses between extra cardiac vessels and the coronary arteries. Acta Med Scand 1968, suppl. 485:5–25.
411. Moffat DB. The embryology of the arteries of the brain. Ann R Coll Surg Engl 1962;30:368–82.
412. Mojab K, Ghosh BC, Moss GS. Bilateral subclavian and axillary artery anomaly. Am Surg 1976;42:585–8.
413. Molsen HP, Winkelmann H. Persistenz embryonaler Kiemenbogenarterien in Verbindung mit weiteren Gefäßanomalien – Übersicht über 14 zerebrale Primitivarterien. Radiol Diagn 1980;21:597–603.
414. Molz G. Abnormer Abgang der Arteriae subclaviae: Häufigeres Vorkommen beim weiblichen Geschlecht. Basic Res Cardiol 1976;71:420–7.
415. Molz G, Burri B. Aberrant subclavian artery (Arteria lusoria): sex differences in the prevalence of various forms of the malformation. Evaluation of 1378 observations. Virchows Arch (Pathol Anat) 1978;380:303–15.
416. Moncada R, Shannon M, Miller R, White H, Friedman J, Shuford WH. The cervical aortic arch. Am J Roentgenol 1975; 125:591–601.
417. Moore TS, Morris JL. Aortic arch vessel anomalies associated with persistent trigeminal artery. Am J Roentgen 1979;133: 309–11.
418. Morettin LB. Coronary arteriography: uncommon observations. Radiol Clin North Am 1976;14:189–208.
419. Mori K, Onoe T, Ooka T. Three main coronary arteries to pulmonary artery fistula. Jpn Circ J 1981;45:209–12.
420. Morris ED, Moffat DB. Abnormal origin of the basilar artery from the cervical part of the internal carotid and its embryological significance. Anat Rec 1956;125:701–11.
421. Morris T, Ruttley M. Left cervical aortic arch associated with aortic aneurysm. Br Heart J 1978;40:87–90.
422. Mozersky DJ, Buckley CJ, Hagord CO, Capps WF, Dannemiller FJ. Ultrasonic evaluation of the palmar circulation: a useful adjunct to radial artery cannulation. Am J Surg 1973;126:810–2.
423. Müller JHA. Doppelung der Arteria femoralis. Fortschr Geb Röntgenstr Nuklearmed 1967;106:152–3.
424. Mundth ED. Surgical anatomy of the coronary arteries. In: Glenn WWL, Bane AE, Geha AS, Hammond GL, Lakes H, eds. Thoracic and cardiovascular surgery. 4th ed. Norwalk: Appleton Century Crofts, 1983;1381:91.
425. Murphy DA, Roy DL, Sohal M, Chandler BM. Anomalous origin of left main coronary artery from anterior sinus of valsalva with myocardial infarction. J Thorac Cardiovasc Surg 1978;75:282–5.
426. Nadas AS, Gamboa R, Hugenholtz PG. Anomalous left coronary artery originating from the pulmonary artery. Report of two surgically treated cases with a proposal of hemodynamic and therapeutic classification. Circulation 1964;29:167–75.
427. Nager GT, Nager M. The arteries of the human middle ear, with particular regard to the blood supply of the auditory ossicles. Ann Otol Rhinol Laryngol 1953;62:923–49.
428. Naidich JB, Naidich TP, Sprayregen S, Hyman RA, Pudlowski RM, Stein HL. The origin of the left gastric artery. Radiology 1978;126:623–6.
429. Naidich TP, Kricheff II, George AE, Lin JP. The normal anterior inferior cerebellar artery. Radiology 1976;119:355–73.
430. Nakamura N. Zur Anatomie der Bronchialarterien. Anat Anz 1924;58:508–17.

431. Narverud G, Myhre HO. Congenital hypoplasia of the lower limb arteries. A report of two cases. Scand J Thorac Cardiovasc Surg 1974;8:70–2.
432. Nath PH, Castaneda-Zuniga W, Zollikofer C, et al. Isolation of a subclavian artery. Am J Roentgenol 1981;137:683–8.
433. Nathan H, Rubinstein Z, Bogart B. Accessory internal thoracic artery. Anat Clin 1982;3:333–7.
434. Nathan H, Tobias PV, Sellsted MD. An unusual case of right and left testicular arteries arching over the left renal vein. Br J Urol 1976;48:135–8.
435. Navarro JAC, Filho JLT, Zorzetto NL. Anatomy of the maxillary artery into the pterygomaxillopalatine Fossa. Anat Anz 1982;152:413–33.
435 a. Nebesar RA, Kornblith PL, Pollard JJ, Michels NA. Celiac and superior mesenteric arteries. A correlation of angiograms and dissections. Boston, Little Brown Co., 1969.
436. Neimann J-L, Ethevenot G, Cuilliere M, Cherrier F. Variations de distribution des artères coronaires. (À propos de 3000 coronarographies) Bull Assoc Anat (Nancy) 1976;60:769–78.
437. Newton TH, Mani RL. The vertebral artery. In: Newton TH, Potts DG, eds. Radiology of the skull and brain. Angiography. Vol 2, Book 2. St. Louis: Mosby, 1974:1659–73.
438. Neye-Bock S, Fellows KE. Aortic arch interruption in infancy: Radio- and angiographic features. Am J Röntgenol 1980;135:1005–10.
439. Nikolić V. O kolateralnim granama gornjeg dijela zajedničke karotide. Rad Med Fak Zagrebu 1961;9:269–78.
440. Nissen R. Operationen am Ösophagus. Stuttgart: Thieme, 1954.
441. Nitschke T, Preuss F. Die Hauptäste der A. iliaca int. bei Mensch und Haussäugetieren in vergleichend-anatomisch häufigster Reihenfolge. Anat Anz 1971;128:439–53.
442. Nitzsche H. Kombination einer seltenen Aortenbogenmißbildung (Cervical aortic arch) mit einer A. lusoria. Fortschr Röntgenstr 1980;132:93–4.
443. Nizankowski C. Seltener Fall eines gemeinsamen Stammes der Schilddrüsenzungengesichtsschlagader beim Menschen (Truncus thyrolinguofacialis). Anat Anz 1972;130:530–4.
444. Noeske K. Über die arterielle Versorgung des menschlichen Rückenmarkes. Gegenbaurs Morph Jahrb 1958;99:455–97.
445. Nordmark L. Angiography of the testicular artery. I Method of examination. Acta Radiol Diagn 1977;18:25–32.
446. Norwood WI, Stellin GJ. Aortic atresia with interrupted aortic arch: Reparative operation. J Thorac Cardiovasc Surg 1981;81:239–44.
447. Notkovich H. Testicular artery arching over renal vein: Clinical and pathological considerations with special reference to varicocele. Br J Urol 1955;27:267–71.
448. Notkovich H. Variations of the testicular and ovarian arteries in relation to the renal pedicle. Surg Gynec Obstet 1956;103:487–95.
449. Nowak M. Variations of the arteries supplying the liver in man. Folia Morphol (Warszawa) 1966;25:263–70.
450. Nowak M. Variation of the celiac trunk in man. Folia Morphol (Warszawa) 1968;27:55–74.
451. Obayashi T, Furuse M. The proatlantal intersegmental artery. A case report and review of the literature. Arch Neurol 1980;37:387–9.
452. Ödman P. Percutaneous selective angiography of the coeliac artery. Acta Radiol (Stockh) 1958;159 (suppl):1–168.
453. Ödman P. Percutaneous selective angiography of the superior mesenteric artery. Acta Radiol (Stockh) 1959;51:25–32.
454. Ödman P. Pancreatic angiography. In: Abrams HL, ed. Angiography, Bd 11. Boston: Little & Brown, 1961;627–38.
455. Ogden JA. Congenital anomalies of the coronary arteries. Am J Cardiol 1970;25:474–9.
456. Okada S, Ohta Y, Shimizu T, Nakamura M, Yaso K. A rare anomalous case of absence of the celiac trunk – the left gastric, the splenic and the common hepatic arteries arose from the abdominal aorta independently. Okajimas Folia Anat Jpn 1983;60:65–72.
457. O'Rahilly R, Debson H, King TS. Subclavian origin of bronchial arteries. Anat Rec 1950;108:227–38.
458. Otulakowski B, Woźniak W. A case of origin of the testicular artery from the renal artery and course of the testicular artery behind the inferior vena cava. Folia Morphol (Warszawa) 1975;34:349–51.
459. Ozaki T, Handa H, Tomimoto K, Hazama F. Anatomical variations of the arterial system of the base of the brain. Arch Jap Chir 1977;46:3–17.
460. Pabst R, Lippert H. Beiderseitiges Vorkommen von A. brachialis superficialis, A. ulnaris superficialis und A. mediana. Anat Anz 1968;123:223–6.
461. Páč L, Hamplová M, Pelcová O. An atypical case of arising of some parietal branches of

the arteria iliaca interna in man. Anat Anz 1977;141:450–4.
462. Padget OH. The development of the cranial arteries in the human embryo. Contrib Embryol Carnegie Inst 1948;32:205–61.
463. Page HL, Engel HJ, Campbell WB, Thomas CS. Anomalous origin of the left circumflex coronary artery. Recognition, angiographic demonstration and clinical significance. Circulation 1974;50:768–73.
464. Palmer FJ. Anomalous origin of the left vertebral artery and its significance in selective femoro-cerebral catheterisation. Australas Radiol 1976;20:225–8.
465. Palmer FJ. Origin of the right vertebral artery from the right common carotid artery: angiographic demonstration of three cases. Brit J Radiol 1977;50:185–7.
466. Palmer FJ, Philips RL. Persistent proatlantal artery arising from the common carotid bifurcation. Aust Radiol 1978;22:226–8.
467. Panek-Mikula J, Górski KA. Developmental and clinical aspects of altered course of the left testicular artery in man. Folia Morphol (Warszawa) 1977;36:323–8.
468. Papadatos D. Trois observations anatomiques sur l'origine de l'arteria subclavia dextra comme dernière branche de l'arcus aortae (arteria lusoria). Anat Anz 1976: 140:100–17.
469. Park SC, Siewers RD, Neches WH, Lenox CC, Zuberbuhler JR. Left aortic arch with right descending aorta and right ligamentum arteriosum. A rare form of vascular ring. J Thorac Cardiovasc Surg 1976; 71:779–84.
470. Paulin S. Coronary angiography. A technical, anatomic and clinical study. Acta Radiol (Diagn) (Stockh) 1964;233 (suppl):84.
471. Paulsen S, Venter M. Anatomical variations of the coronary arteries and origin of blood supply to sinoauricular node determined on the basis of postmortem coronary angiography. Acta Pathol Microbiol Scand (A) 1973;81:784–90.
472. Pedal I. Aortale Ursprungsanomalie einer Koronararterie. Ursache für den plötzlichen Herztod eines 10jährigen Mädchens. Dtsch Med Wochenschr 1976;101:1601–4.
473. Penteado CV. Anatomosurgical study of the superficial and deep circumflex iliac arteries. Anat Clin 1983;5:125–7.
474. Perret W. Die Rechtsprechung bei versehentlicher intraarterieller Injektion anläßlich einer intravenösen Einspritzung in die Ellenbeuge. (Schlußwort) Med Klin 1965;60:2069–70.
475. Perry LW, Scott LP. Anomalous left coronary artery from pulmonary artery: report of 11 cases: review of indications for and results of surgery. Circulation 1970; 41:1043–52.
476. Petelenz T. Right coronary artery syndrome and extracoronary vascularization of myocardium in man. Folia Cardiol (Milano) 1966;25:143.
477. Petrén T. Die Arterien und Venen des Duodenums und des Pankreaskopfes beim Menschen. Z Anat Entwicklungsgesch 1929;90:234–77.
478. Pfefferkorn JR, Löser H, Pech G, Toussaint R, Hilgenberg F. Absent pulmonary artery. A hint to its embryogenesis. Pediatr Cardiol 1982;3:283–6.
479. Pick JW, Anson BJ. The inferior phrenic artery: Origin and suprarenal branches. Anat Rec 1940;78:413–27.
480. Pick JW, Anson BJ, Ashley, FL. The origin of the obturator artery: a study of 640 body halves. Am J Anat 1942;70:317–43.
481. Pierpont ME, Zollikofer CL, Moller JH, Edwards JE. Interruption of the aortic arch with right descending aorta. A rare condition and a cause of bronchial compression. Pediatr Cardiol 1982;2:153–9.
482. Pierson HH. Seven arterial anomalies of the human leg and foot. Anat Rec 1925; 30:139–145.
483. Pierson JM. The arterial blood supply of the pancreas. Surg Gynecol Obstet 1943; 77:426–32.
484. Pillet J, Cronier P, Mercier P, Chevalier JM. The ischiopopliteal arterial trunk. A report of two cases. Anat Clin 1982;3:329–31.
485. Pillet J, Enon B, Cronier P, Mercier P, Moreau P, Chevalier JM. Role of the mesenteric arteries in the arterial blood supply to the lower limbs. Anat Clin 1982;4:307–14.
486. Pilz E. Zwischenfälle bei intraarteriellen Injektionen. Med Klin 1965;60:2018–21.
487. Pinstein ML, Gerald B. Anomalous communication of the external carotid and vertebral arteries. Persistence of the proatlantal artery. Radiology 1976;118:626.
488. Pirker E, Schmidberger H. Die Arteria ischiadica. Eine seltene Gefäßvariante. Fortschr Geb Röntgenstr Nuklearmed 1972; 116:434–7.
489. Piscol K. Die Blutversorgung des Rückenmarks und ihre klinische Relevanz. Berlin, Heidelberg, New York: Springer Verlag, 1972.
490. Platzer W. Die Variabilität der Arteria carotis interna im Sinus cavernosus in Beziehung zur Variabilität der Schädelbasis. Gegenbaurs Morphol Jahrb 1957; 98:227–43.

491. Plengvanit U, Chearanai O, Sindhvananda K, Damrongsak S, Viranuvatti V. Collateral arterial blood supply of the liver after hepatic artery ligation, angiographic study of twenty patients. Ann Surg 1972; 175:105–10.
492. Poisel S, Golth D. Zur Variabilität der großen Arterien im Trigonum caroticum. Wien Med Wochenschr 1974;124:229–32.
493. Poisel S, Spängler HP. Über aberrante und akzessorische Nierenarterien bei Nieren in typischer Lage. Anat Anz 1969;124:244–59.
494. Poisel S, Spängler HP. Die Verästelungstypen der A. renalis im Hinblick auf die arterielle Blutversorgung des Parenchyms der Niere. Ein Beitrag zum Problem der sogenannten Nierensegmente. Acta Anat 1970;76:516–29.
495. Pokieser H. Angiographie der abdominellen Organe. Ergeb. Med. Radiol. 1972;4:1–82.
496. Pool PE, Vogel JHK, Blount SG. Congenital unilateral absence of a pulmonary artery. The importance of flow in pulmonary hypertension. Am J Cardiol 1962;10:706–32.
497. van Praagh R, van Praagh S. Persistent fifth arterial arch in man. Am J Cardiol 1969;24:279–82.
498. Puchades-Orts A, Nombela-Gomez M, Ortuno-Pacheco G. Variation in form of circle of Willis: some anatomical and embryological considerations. Anat Rec 1975; 185:199–24.
499. Puerta CV, Puente JR. Variaciones en la distribucion de la arteria femoral profunda. Estudio de 105 preparaciones quirurgicas. Ann Anat 1982;31:133–8.
500. Quie RG, Novick R, Adams P, Anderson RC, Varco RL. Congenital interruption of the aortic arch. J Pediatr 1959;54:87–92.
501. Radojevic S, Stolic E. Note sur l'artère spermatique interne. Origine haute en rapport avec le pédicale rénale. Bull Assoc Anat (Nancy) 1964;122:314–21.
502. Rajaram PC, Hussain AT, Lakshmikanthan C, Vaidyanathan D, Subramanian N. Type D double aortic arch. Double aortic arch with interruption of its left component proximal to the site of origin of left common carotid artery. Angiology 1983;34:597–603.
503. Rao TS, Sethi PK. Persistent proatlantal artery with carotid-vertebral anastomosis. J Neurosurg 1975;43:499–501.
504. Redman HC, Reuter SR. Angiographic demonstration of surgically important vascular variations. Surg Gynec Obstet 1969; 129:33–39.
505. Rees S. Arterial connections of the lung. The inaugural Keith Jefferson lecture. Clin Radiol 1981;32:1–15.
506. Reid DA, Foster ED, Stubberfield J, Alley RD. Anomalous right subclavian artery arising proximal to a postductal thoracic aortic coarctation. Ann Thorac Surg 1981; 32:85–7.
507. Reimann B, Lierse W, Schreiber H.-W. Anastomosen zwischen Segmentarterien der Leber und phrenico-hepatische arterio-arterielle Anastomosen. Langenbecks Arch Chir 1983;359:81–92.
508. Resche F, Resche-Perrin I, Robert R, de Kersaint-Gilly A, Duveau D, Lajat Y. L'artère hypoglosse. Rapport d'un nouveau cas – Revue de la littérature. J Neuroradiol 1980;7:27–43.
509. Reschke H. Die selektive Angiographie der A. phrenica ascendens. Fortschr Geb Röntgenstr Nuklearmed 1967;107:200–5.
510. Reuter SR. Superselective pancreatic angiography. Radiology 1969;92:74–85.
511. Reynolds AF, Stovring J, Turner PT. Persistent otic artery. Surg Neurol 1980; 13:115–7.
512. Richardson JV, Doty DB, Rossi NP, Ehrenhaft JL. The spectrum of anomalies of aortopulmonary septation. J Thorac Cardiovasc Surg 1979;78:21–7.
513. Richardson JV, Doty DB, Rossi NP, Ehrenhaft JL. Operation for aortic arch anomalies. Ann Thorac Surg 1981;31:426–32.
514. Riggs HE, Rupp C. Variations in form of circle of Willis. Arch Neurol 1963;8:24–30.
515. Riva D. Considerazioni anatomo-chirurgiche sul tronco celiaco ed i suoi rami. Minerva Chir 1967;22:570–9.
516. Roberts LK, Gerald B. Absence of both common carotid arteries. Am J Roentgenol 1978;130:981–2.
517. Roberts WC, Siegel RJ, Zipes DP. Origin of the right coronary artery from the left sinus of valsalva and its functional consequences: analysis of 10 necropsy patients. Am J Cardiol 1982;49:863–8.
518. Roberts WH, Engen PC, Mitchell DA. When the marginal artery is not marginal. Anat Anz 1984;155:269–72.
519. Roberts WH, Krishingner GL. Comparative study of human internal iliac artery based on Adachi classification. Anat Rec 1967;158:191–6.
520. Robertson PW, Hull DH, Klidjian A, Dyson ML. Renal artery anomalies and hypertension. A study of 340 patients. Am Heart J 1967;73:296–307.
521. Robillard GL, Shapiro AL. Variational anatomy of the middle colic artery. Its significance in gastric and colonic surgery. J Int Coll Surg 1947;10:157–69.

522. Robinson LR. An unusual human anterior cerebral artery. J Anat 1959;93:131–3.
523. Rösch J. Arteriography of pancreas. Am J Roentgenol 1965;94:182–93.
524. Rosen IU, Mills DF, Nadel HI, Kaiserman DD. Angiographic demonstration of congenital absence of both internal carotid arteries. J Neurosurg 1975;42:478–82.
525. Rouanet JP, Bruel JM, Lamarque JL, Djoukhadar A. Arteriographic aspects of the main anatomical and physiological arterial digestive collateral circulations. Anat Clin 1979;1:375–82.
526. Rowe GG, Young WP. Anomalous origin of the coronary arteries with special reference to surgical treatment. J Thorac Cardiovasc Surg 1960;39:777–80.
527. Rowe LD, Lowry LD, Keane WM, Fallenjad M. Aortic arch anomalies in adult disorders of deglutition. Ann Otol Rhinol Laryngol 1978;87:498–508.
528. Ruzicka FF, Rossi P. Normal vascular anatomy of the abdominal viscera. Radiol Clin North Am 1970;8:3–29.
529. Sabiston DC, Neill CA, Taussig MB. The direction of blood flow in anomalous left coronary artery arising from the pulmonary artery. Circulation 1960;22:591–7.
530. Sacks JH, Londe SP, Rosenbluth A, Zalis EG, Judkins MP. Left main coronary bypass for aberrant (aortic) intramural left coronary artery. J Thorac Cardiovasc Surg 1977;73:733–7.
531. Sade RM, Rosenthal A, Fellows K, Castañeda AR. Pulmonary artery sling. J Thorac Cardiovasc Surg 1975;69:333–46.
532. Salomon J, Baltazar R, Mower MM, Goldman S. Anomalous origin of the right coronary artery from the pulmonary artery. Am J Med Sci 1981;281:152–6.
533. Saltzmann GF. Patent primitive trigeminal artery studied by cerebral angiography. Acta Radiol (Stockh) 1959;51:329–36.
534. Samra K, Scoville WB, Yaghmai M. Anastomosis of carotid and basilar arteries. Persistant primitive trigeminal artery and hypoglossal artery: report of two cases. J Neurosurg 1969;30:622–5.
535. Sansom JR, Hall CL, Barnes AD, Robinson BHB, Blainey JD. The significance of vascular anomalies in human renal transplantation. Proc Eur Dial Transplant Assoc 1978;15:345–51.
536. Sarto RK, Freckmann N, Böker DK. Related anomalies of origin of left vertebral and left inferior thyroid arteries. Neuroradiology 1980;19:27–30.
537. Sartor K. Die selektive Spinalarteriographie und ihre Bedeutung für Diagnostik und Therapie spinaler Prozesse, insbesondere arteriovenöser Mißbildungen. Röntgen Bl 1977;30:607–15.
538. Schaffer R, Gordon DH, Glanz S. Renal artery originating above celiac axis. NY State J Med 1981;81:1109–11.
539. Schlesinger MJ, Zoll PM, Wessler S. The conus artery: A third coronary artery. Am Heart J 1949;38:823–36.
540. Schmidt H, Beyer L. Das Angiogramm der Schilddrüse (2. Mitteilung). Fortschr Röntgenstr 1968;108:376–81.
541. Schmidt H, Pfingst E. Zur Dopplung der A. vertebralis. Fortschr Geb Röntgenstr Nuklearmed 1973;118:636–40.
542. Schmitt HP. A rare abnormality of the cerebral basilary arterial circle. Eur Neurol 1973;10:75–82.
543. Schober R. Selektive Bronchialisarteriographie. Fortschr Geb Röntgenstr Nuklearmed 1964;101:337–48.
544. Schoenmackers J. Koronararterien. Herzinfarkt. In: Bargmann W, Doerr W, eds. Das Herz des Menschen, 2nd ed. Stuttgart: Thieme, 1963:755–92.
545. Schoenmackers J, Scheunemann H. Angiographische Untersuchungen der A. carotis externa. Dtsch Zahn-, Mund-, Kieferheilk 1956;23:345–60.
546. Schorn J, Stender HS, Voegt H. Untersuchungen über die arterielle Strombahn der Leber. Langenbecks Arch Klin Chir 1957;286:187–217.
547. Schossberger P. Vasculature of the spinal cord: a review. Part I. Anatomy and physiology. Bull Los Angeles Neurol Soc 1974;39:71–97.
548. Schreiber H. Seltene Varietät der A. carotis. Anat Anz 1941;91:269–71.
549. Schubel AL. Abnormer Verlauf der 10. Interkostalarterie in Verbindung mit seltenen Gefäßvarietäten im Bereich der A. coeliaca. Anat Anz 1955;102:279–86.
550. Schwyzer AG, de Garis CF. Three diverse patterns of the arteria brachialis superficialis in man. Anat Rec 1935:63:405–16.
551. Scott HS. Carotid basilar anastomosis – persistent hypoglossal artery. Br J Radiol 1963;36:847–9.
552. Sebastiani M, Romano V, Gerletti D, Greco O. Contributo all'anatomia chirurgica dell'arteria splenica. Le connessioni vascolari arteriose pancreatico-spleniche. Biol Lat 1968;21:621–30.
553. Seeger JF, Hemmer JF. Persistent basilar/middle meningeal artery anastomosis. Radiology 1976;118:367–70.
554. Seidel K. Arteriographische Beobachtung einer seltenen Carotisanomalie. Fortschr

Geb Röntgenstr Nuklearmed 1965;103: 390–1.
555. Seldinger SI. Arteries of the extremities. In: Diethelm L, Olsson O, Strnad F, Vieten H, Zuppinger A, eds. Handbuch Medizinischer Radiologie, Vol. 10, part 3. Berlin: Springer, 1964:400–72.
556. Sengupta RP. Proceedings: Anatomical variations in the origin of the posterior cerebral artery demonstrable by carotid angiography and their significance to the direct surgical treatment of posterior communicating aneurysms. Acta Neurochir (Wien) 1975;31:271–2.
557. Senior HD. Abnormal branching of the human popliteal artery. Am J Anat 1929;44:111–20.
558. Servida E, Mortillaro F. Studio morfologico delle anomalie delle arterie cerebrali (Osservazioni su 200 poligoni di Willis). Folia Hered Pathol (Milano) 1962;11:245–55.
559. Shah AC, Srivastava HC. Bifurcation of the common carotid artery. Indian J Surg 1965;27:662–4.
560. Shakibi JG, Rastan H, Nazarian I, Paydar M, Aryanpour I, Siassi J. Isolated unilateral absence of the pulmonary artery. Review of the world literature and guidelines for surgical repair. Jpn Heart J 1978;19:439–51.
561. Shaner RF. The persisting right sixth aortic arch of mammals, with a note on fetal coarctation. Anat Rec 1956;125:171–84.
562. Shapiro AL, Robillard GL. The esophageal arteries. Their configurational anatomy and variations in relation to surgery. Ann Surg 1950;131:171–85.
563. Sharratt GP, Leanage R, Monro JL, Shinebourne EA. Aortic arch interruption presenting with absence of all limb pulses. Arch Dis Child 1979;54:49–53.
564. Shepherd RM, Kerth WJ, Rosenthal JH. Right cervical aortic arch with left descending aorta. Am J Dis Child 1969;118:642–8.
565. Sher RF, Iskandrian AS, Kimbiris D, Bemis CE. Anomalous origin of the left coronary artery from the right sinus of valsalva. Cathet Cardiovasc Diagn 1978;4:413–7.
566. Shimada Y, Fukuyama U. Testicular and ovarian blood vessels. Acta Anat Nipponica 1975;50:147–50.
567. Shuford WH, Sybers RG. The aortic arch and its malformations. Springfield, Il: Charles C. Thomas, 1974:30–1:54.
568. Shuford WH, Sybers RG, Edwards FK. The three types of right aortic arch. Am J Roentgen 1970;109:67–74.
569. Shuford WH, Sybers RG, Weens HS. The angiographic features of double aortic arch. Am J Roentgenol 1972;116:125–40.
570. Siddharth P, Smith NL. An anatomic basis to prevent ischemia of the colon during operations upon the aorta. Surg Gynecol Obstet 1981;153:71–3.
571. Sierocinski W, Kling A, Michniewicz-Nowak M. The occurrence and the course of arteries vascularizing the cecum and the colon in man. Folia Morphol (Warszawa) 1981;40:149–57.
572. Sierocinski W, Kling A, Michniewicz-Nowak M. The occurrence and the course of "the marginal artery" as the anastomosis of arteries vascularizing the cecum and the colon in man. Folia Morphol (Warszawa) 1981;40:167–79.
573. Sierocinski W, Michniewicz-Nowak M. Variation of the arising and the course of arteries vascularizing the transverse colon in man. Folia Morphol (Warszawa) 1980; 39:271–81.
574. Singer E. Embryological pattern persisting in the arteries of the arm. Anat Rec 1933;55:403–9.
575. Skopakoff C. Über die Variabilität der Ab- und Verzweigung der A. brachialis superficialis. Anat Anz 1959;106:356–68.
576. Skopakoff C. Über die Variabilität der Abzweigung der Hauptäste der Aa. axillaris et brachialis. Anat Anz 1959;107:284–304.
577. Smith JC. Review of single coronary artery with report of two cases. Circulation 1950;1:1168–75.
578. Smith SC, Adams DF, Herman MV, Paulin S. Coronary to bronchial anastomoses: an in vivo demonstration by selective coronary arteriography. Radiology 1972;104:289–90.
579. Smith SD, Benton RS. A rare origin of the superior thyroid artery. Acta Anat 1978;101:91–3.
580. Smith-Agreda V. Posibilidades genotipicas en ciertas ramos de distribución de las arterias del miembro superior. An Anat (Granada) 1959;8:53–74.
581. Somerville J, Saravalli O, Ross D. Complex pulmonary atresia with congenital systemic collaterals. Classification and management. Arch Mal Coeur 1978;71:322–8.
582. Sonneland J, Anson BJ, Beaton LE. Surgical anatomy of the arterial supply to the colon from the superior mesenteric artery based on a study of 600 specimens. Surg Gynecol Obstet 1958;106:385–98.
583. Soutoul JH, Isabellon C. A propos du niveau d'origine de l'artère mésentérique inférieure. Arch Anat Pathol (Paris) 1960;8:66–9.
584. Spalteholz W. Die Arterien der Herzwand. Leipzig: Hirzel, 1924.

585. Spindola-Franco H, Weisel A, Delman AJ. Pulmonary steal syndrome: an unusual case of coronary-bronchial pulmonary artery communication. Radiology 1978;126:25–7.
586. Ssoson-Jaroschewitsch AJ. Zur chirurgischen Anatomie des Milzhilus. Z Anat Entwicklungsgesch 1927;84:218–37.
587. Stampfel G, Haertel M. Aortaler Ursprung der Arteria ileocolica – oder Arteria mesenterica media? Fortschr Geb Röntgenstr Nuklearmed 1976;124:495–6.
588. Stanfield JP. The blood supply of the human pituitary gland. J Anat 1960; 94:257–73.
589. Starck D. Embryologie, 2nd ed. Stuttgart: Thieme 1965.
590. Steffen TN. Vascular anomalies of the middle ear. Laryngoscope 1968;78:171–97.
591. Stein MG, Christie PW. Asymptomatic isolated left subclavian artery with right aortic arch. A case report. S Afr Med J 1982;62:333–5.
592. Stern J, Correli JW, Bryan N. Persistent hypoglossal artery and persistent trigeminal artery presenting with posterior fossa transient ischemic attacks. J Neurosurg 1978;49:614–9.
593. Steward JA, Rankin FW. Blood supply of the large intestine, its surgical considerations. Arch Surg 1933;26:843–91.
594. Stulberg HK. Selective hepatic arteriography. Normal anatomy, anatomic variations and pathological conditions. Radiology 1965;85:36–55.
595. Suenaga Y, Nakamura R, Inoue K, Terashima T, Takeda M, Inoue Y. Two cases of a lieno-mesenteric trunk. Okajimas Folia Anat Jpn 1982;58:1249–58.
596. Sundgren, R. Selective angiography of the left gastric artery. Acta Radiol (Stockh) 1970;299 (suppl).
597. Sutton D, Davies ER. Arch aortography and cerebrovascular insufficiency. Clin Radiol 1966;17:330–45.
598. Suzuki K, Prates JC, Didio LJA. Incidence and surgical importance of the posterior gastric artery. Ann Surg 1978;187:134–6.
599. Suzuki S, Kuwabara Y, Hatano R, Iwai T. Duplicate origin of left vertebral artery. Neuroradiology 1978;15:27–9.
600. Suzuki T, Nakayasu A, Kawabe K, Takeda H, Honjo I. Surgical significance of anatomic variations of the hepatic artery. Am J Surg 1971;122:505–12.
601. Swigart LVL, Siekert RG, Hambley WC, Anson BJ. The esophageal arteries. An anatomic study of 150 specimens. Surg Gynecol Obstet 1950;90:234–43.
602. Sykes D. The arterial supply of the human kidney with special reference to accessory renal arteries. Br J Surg 1963;50:368–74.
603. Sylla S, Papasian P, Dintimille H, Argenson C. Repérage du nerf récurrent et de l'artère thyroïdienne inférieure d'après 68 dissections chez l'Africain à Dakar. Bull Soc Med Afr Noire Lang Franc 1976;21:30–9.
604. Sylla S, Papasian PP, Johnson E, Argenson C. Le tronc thyrobicervico-scapulaire. Classification d'après 120 dissections chez l'Africain à Dakar. Bull Soc Med Afr Noire Lang Franc 1978;23:7–13.
605. Takemura A, Masomoto M, Mori T. A case of the right subclavian artery as the last branch of the aortic arch. Okajimas Folia Anat Jpn 1979;56:317–28.
606. Tan WS, Spigos DG. Right vertebral artery originating from the right common carotid. Radiologe 1979;19:155–6.
607. Teal JS, Rumbaugh CL, Bergeron RT, Scanlan RL, Segall HD. Persistent carotid superior cerebellar artery anastomosis: a variant of persistent trigeminal artery. Radiology 1972;103:335–41.
608. Teal JS, Rumbaugh CL, Bergeron RT, Segall HD. Anomalies of the middle cerebral artery: accessory artery, duplication and early bifurcation. Am J Roentgenol 1973;118:567–75.
609. Teal JS, Rumbaugh CL, Segall HD, Bergeron RT. Anomalous branches of the internal carotid artery. Radiology 1973;106:567–73.
610. Tedeschi CG, Helpern MM. Heterotopic origin of both coronary arteries from the pulmonary artery: review of literature and report of a case not complicated by associated defects. Pediatrics 1954;14:53–7.
611. Thomsen G, Vesterdal J, Husfeldt E. Abnormalities of great vessels. A case of double aortic arch and two cases of anomalous right subclavian artery. Acta Otolaryngol (Stockh.) 1953;109:182–8.
612. Tiraboschi R, Crupi G, Villani M, et al. Anelli vascolari sintomatici nella prima infanzia. G Ital Cardiol 1981;11:68–78.
613. Tischendorf F. Zum Problem der Milzarterie. II. Eine abnorm stark und unregelmäßig geschlängelte A. lienalis mit zweiteiligem Truncus coeliacus. Anat Anz 1973; 134:108–19.
614. Tisnado J, Amendola MA, Beachley MC. Renal artery originating from the inferior mesenteric artery. Br J Radiol 1979; 52:752–4.
615. Tonkin IL, Elliott LP, Bargeron LM. Concomitant axial cineangiography and barium esophagography in the evaluation of vascular rings. Radiology 1980;135:69–76.

616. Torre E de la. Study of persistent primitive maxillary artery in human fetus: some homologies of cranial arteries in man and dog. Am J Anat 1960; 106: 185–95.
617. Tramier D, Argême M, Huguet JF, Juhan C. Radiological anatomy of the internal pudendal artery (a. pudenda interna) in the male. Anat Clin 1981; 3: 195–200.
618. Trivellato M, Angelini P, Leachman RD. Variations in coronary artery anatomy: normal versus abnormal. Cardiovasc Dis Bull Texas Heart Inst 1980; 7: 357–70.
619. Trivellato M, Angelini P, Lufschanowski R, Krajcer Z, Leachman R. Classificazione e significato clinico delle anomalie del sistema arterioso coronarico. G Ital Cardiol 1980; 10: 346–55.
620. Trotter M. The level of termination of the popliteal artery in the white and the negro. Am J Phys Anthropol 1940; 27: 109–18.
621. Trotter M, Curtis GH, Westbrook ER. The origins of branches of the axillary artery in whites and in American negroes. Anat Rec 1930; 46: 133–7.
622. Tsai FY, Mahon J, Woodruff JV, Roach JF. Congenital absence of bilateral vertebral arteries with occipital-basilar anastomosis. Am J Roentgenol 1975; 124: 281–6.
623. Tsukamoto S, Hori Y, Utsumi S, Tanigake T, Horiike N, Otani R. Proatlantal intersegmental artery with absence of bilateral vertebral arteries. J Neurosurg 1981; 54: 122–4.
624. Tucker BL, Meyer BW, Lindensmith GG, Stiles QR, Jones JC. Congenital aortic vascular ring. Arch Surg 1969; 99: 521–3.
625. Turnbull I. Agenesis of the internal carotid artery. Neurology (NY) 1962; 12: 588–90.
626. Turnbull IM. Blood supply of the spinal cord: normal and pathological considerations. Clin Neurosurg 1973; 20: 56–84.
627. Turner RS. The angle of origin of the ulnar artery. Anat Rec 1959; 134: 761–8.
628. Umek H. Zur Relevanz anatomischer Varianten im Oberbaucharterienbereich. Röntgenpraxis 1973; 26: 101–6.
629. Vancov V. Une variété extrêmement complexe des artères du membre supérieur chez un foetus humain. Anat Anz 1961; 109: 400–4.
630. Vandamme JPJ. A revaluation of hepatic and cystic arteries. Acta Anat 1969; 73: 192–209.
631. Vandamme JPJ, Bonte J. A new look at the blood supply of the ileocolic angle. Acta Anat 1982; 113: 1–14.
632. Vandamme JPJ, Bonte J. Arcus paracolicus. Acta Anat 1984; 118: 50–3.
633. Vandamme JPJ, Bonte J, van der Schueren G. A revaluation of hepatic and cystic arteries. The importance of the aberrant hepatic branches. Acta Anat 1969: 73; 192–209.
634. Vandamme JPJ, Bonte J, van der Schueren G. Re-evaluation of the colic irrigation from the inferior mesenteric artery. Acta Anat 1982; 112: 18–30.
635. Vandamme JPJ, van der Schueren G. Reevaluation of the colic irrigation from the superior mesenteric artery. Acta Anat. 1976; 95: 578–88.
636. Varro J, Laszlo H, Varga G. Anatomy of the hand arteries based on angiographic studies. Magyar Traumat Orthop 1978; 21: 127–34.
637. Ventemiglia R, Oglietti J, Wukasch DC, Hallman GL, Cooley DA. Interruption of the aortic arch. Surgical considerations. J Thorac Cardiovasc Surg 1976; 72: 235–42.
638. Vincent J, Lacquet LK, Kuijpers PJ, Brom AG. Congenitale vasculaire ringen door aortabooganomalieën. Acta Chir Belg 1975; 74: 305–16.
639. Vitek JJ, Reaves P. Thoracic bifurcation of the common carotid artery. Neuroradiology 1973; 5: 133–9.
640. Vlodaver Z, Amplatz HB, Burchell HH, Edwards JE. Coronary heart disease. Clinical, angiographic and pathologic profiles. New York: Springer, 1976.
641. Vlodaver Z, Neufeld HN, Edwards JE. Coronary arterial variations in the normal heart and in congenital heart disease. New York: Academic Press, 1975.
642. Vogelsang H. Über eine angiographisch selten nachgewiesene Anastomose zwischen dem Art. carotis interna- und dem Art. carotis externa-Kreislauf. Nervenarzt 1961; 32: 518–20.
643. Vogelsang H. Angiographische Untersuchungen von Wirbelsäule, Spinalkanal und Rückenmark. In: Diethelm L, Heuck F, Olsson O, Strnad F, Vieten H, Zuppinger A, eds. Handbuch der medizinischen Radiologie. Berlin, Heidelberg: Springer Verlag, 1981; 177–268.
644. Volkmann J. Zur chirurgischen Anatomie der Milzgefäße. Zentralblatt Chir 1923; 50: 436–8.
645. Voss H. Eine seltene Varietät des Truncus coeliacus der Bauchaorta: Arteria pancreatica. Anat Anz 1959; 107: 329–31.
646. Wachtler F. Eine seltene Varietät im Gefäß-Nervenbündel der Axilla. Anat Anz 1978; 143: 450–5.
647. Waller BJ. Five coronary ostia: duplicate left anterior descending and right comis coronary arteries. Am J Cardiol 1983; 51: 1562.
648. Wallraff J, Roszel A. Hochgradige Hypoplasie der Arteria iliaca communis und der

Arteria umbilicalis. Anat Anz 1972; 132:101–13.
649. Wankoff W. Die Frage der Typenbestimmung der A. brachialis superficialis im Lichte einer neuen Varietät, die beim menschlichen Fötus beobachtet wurde. Anat Anz 1961;110:265–9.
650. Wankoff W. Über einige Gesetzmäßigkeiten bei der Variabilität der Arterien der oberen Extremitäten. Anat Anz 1962;111:216–40.
651. Wappenschmidt J. Über eine seltene Verlaufsanomalie der proximalen Carotis communis und deren klinische Bedeutung. Fortschr Geb Röntgenstr Nuklearmed 1967;107:145–7.
652. Watanabe T, Togo M. Accessory middle cerebral artery. Report of four cases. J Neurosurg 1974;41:248–51.
653. Weatherby HT. Unusual variation of the ulnar artery. Anat Rec 1956;124:245–8.
654. Webb WR, Gamsu G, Speckman JM, Kaiser JA, Federle MP, Lipton MJ. CT demonstration of mediastinal aortic arch anomalies. J Comp Ass Tomogr 1982;6:445–51.
555. Weißleder H, Baumeister L, Fischer P, Renemann H. Die selektive Darstellung der Arteriae coeliaca und mesenterica superior in der abdominalen Diagnostik. Fortschr Geb Röntgenstr Nuklearmed 1966;104:137–49.
656. Wesselhoeft H, Fawcett JS, Johnson AL. Anomalous origin of the left coronary artery from the pulmonary trunk: its clinical spectrum, pathology and pathophysiology, based on a review of 140 cases with 7 further cases. Circulation 1968;38:403–25.
657. White NK, Edwards JE. Anomalies of the coronary arteries. Report of four cases. Arch Pathol Lab Med 1948;45:766–71.
658. Wicke L, Spängler HP, Dimopoulos J, Firbas W, Olbert F. Zur Dignität der Variationen der Nierenarterienabgänge im Angiogramm. Anat Anz 1974;135:140–50.
659. Widmer O. Die Rectalarterien des Menschen. Eintritt, Kaliber, Verteilungsart, Anastomosen und Versorgungsgebiete. Z Anat Entwickl-Gesch 1955;118:398–416.
660. Wiedner F, Schreyer H. Seltene Anomalien zerebraler Gefäße (einseitige Aplasie der A. carotis interna, Persistenz der A. hypoglossica). Fortschr Geb Röntgenstr Nuklearmed 1976;124:245–9.
661. Williams GD, Edmunds HW. Variations in arrangements of branches arising from aortic arch in American whites and negroes. Anat Rec 1935;62:139–46.
662. Williams LR, Flanigan DP, O'Connor RJA, Schuler JJ. Persistent sciatic artery: Clinical aspects and operative management. Am J Surg 1983;145:687–93.
663. Wilson CL, Dlabal PW, Holeyfield RW, Akins CW, Knauf DG. Anomalous origin of left coronary artery from pulmonary artery. Case report and review of literature concerning teenagers and adults. J Thorac Cardiovasc Surg 1977;73:887–93.
664. Wilson PM. An unusual variation of the coronary arteries. Anat Anz 1965; 116:299–302.
665. Wirtanen GW, Kaude JV. Inferior phrenic artery collateralization in hepatic artery occlusion. Am J Roentgenol 1973;117:615–9.
666. Wise BL, Palubinskas AJ. Persistent trigeminal artery (carotid-basilar anastomosis). J Neurosurg 1964;21:199–206.
667. Wójtowicz J, Zierhoffer S. Ein kasuistischer Beitrag zur klinischen Bedeutung der Entwicklungsanomalie der Arteria ovarica. Fortschr Geb Röntgenstr Nuklearmed 1967;107:152–4.
668. Woodburne RT, Olsen LL. The arteries of the pancreas. Anat Rec 1951;111:255–72.
669. Wozniak W, Kiersz A, Wawrzyniak S. The question of the renal arterial segments. Anat Anz 1972;132:332–40.
670. Wright CW. Aberrant blood vessels and nerves in a case of left-sided vermiform appendix, cecum, and anomalous colon. Anat Rec 1959;133:187–202.
671. Wyler F, Graedel E, Rutishauser M, Klimm C, Hasse J. Successful palliation by means of a bovine artery graft in a 4 day old infant with type B interruption of aortic arch and right descending aorta. Thorac Cardiovasc Surg 1980;28:57–60.
672. Xuereb GP, Prichard MML, Daniel PM. The arterial supply and venous drainage of the human hypophysis cerebri. Q J Exp Physiol 1954;39:199–217.
673. Yamaki S, Suzuki Y, Kagawa Y, et al. Histologic findings of ductus arteriosus and pulmonary vascular disease in interrupted aortic arch. Tohoku J Exp Med 1982;136:291–8.
674. Yoshida Y, Emmanouilides GC, Nelson RJ, Baylen BG, Criley JM. Anomalous origin of the left coronary artery from the pulmonary artery: A case report with remarkable improvement of myocardial function following subclavian artery – coronary artery anastomosis. Cathet Cardiovasc Diagn 1980;6:293–303.
675. Youngson GG, Taylor B, Rankin R, Heimbecker RO. Persistent sciatic artery: A case report. Can J Surg 1980;23:466–7.
676. Zahn DGH, Goerttler K. Über die Sklerose der Eingeweidearterien. Pathomorphologische Befunde, Lokalisation und Häufigkeit. Arch Kreislaufforsch 1971;64:235–72.

Index

Main references are given in bold types: the term artery, branch, ramus, etc. is always omitted

GPSR Compliance

The European Union's (EU) General Product Safety Regulation (GPSR) is a set of rules that requires consumer products to be safe and our obligations to ensure this.

If you have any concerns about our products, you can contact us on ProductSafety@springernature.com

In case Publisher is established outside the EU, the EU authorized representative is:

Springer Nature Customer Service Center GmbH
Europaplatz 3
69115 Heidelberg, Germany

Batch number: 10683525

Printed by Printforce, the Netherlands